BIOPSY OF RELIGIONS

Abhijit Naskar, bestselling author and one of the world's celebrated neuroscientists, takes us on a tour of the vivid and mysterious realm of the human mind in all his scientific works with his peerless explanatory ways. Here in this book, through close observation of the fascinating neurological circuits of the human brain, he shows us the biological origin of all the conflicts concerning religion. He enables us to tap into our own human consciousness and find absolution from all those conflicts.

I0464739

BIOPSY
OF
RELIGIONS

NEUROANALYSIS TOWARDS
UNIVERSAL TOLERANCE

ABHIJIT
NASKAR

An Amazon Publishing Company, 1st Edition, 2016

Printed in United States of America

ISBN-13: 978-1523953028

To know that we know what we know, and to know that we do not know what we do not know, that is true knowledge.

-     Nicolaus Copernicus (1473-1543)

Dedicated to my dearly beloved Lizi.

One nice evening we were having a discussion on the Syrian conflict. Being an expert in International Relations, she was telling me all about the history of the Middle-East. My book Autobiography of God: Biopsy of A Cognitive Reality had just been released. So, I was explaining to her how people jump to the conclusion, that religion is the cause of terrorism, without any deeper understanding of the matter. Suddenly out of the blue, she said *"why don't you write a book on this!"*

And those were the words that quite instantly overwhelmed my entire brain with an enthralling storm of neural firings, that didn't even let me have a sound sleep until I finished the book. If she hadn't told me those exact words at that exact moment, I might not have written this book, just yet.

CONTENTS

AUTHOR'S NOTE

Sisters and Brothers from Mother Nature,

It fills my heart with unspeakable pain-stricken joy to rise in response to the 21st century conflicts related to religions. But, the very fact that such religious conflicts still exist in this modern age, is proof that no matter how much we advance in the path of technological development, deep down we'll always remain Apes. And the most delightful evidence of such primitiveness can be seen in those religious fundamentalists who do not let go of a single opportunity to boast their intellectual prowess. To the untrained eye these primeval creatures often seem to be highly intelligent human scholars in the field of religious studies.

Whereas the reality is, one who says, *"my religion is the only true religion, and all others are false"* or *"my God is the true God, and all others are fake"*, is neither religious nor a Homo sapiens, in a manner of speaking; could be an early Australopithecine. When the fundamentalists boast about their own beliefs while denying all other beliefs, they probably look like children in the pre-school, who keep shouting *"my daddy is the best".* Fundamentalism is dangerous. It drags the whole human civilization back to the stone-age. As a neuroscientist, I take full responsibility in telling you that fundamentalism brings nothing but chaos onto the world.

I am simply a scientist who investigates the neurological underpinnings of all kinds of human experiences including the religious belief systems. And as a neuroscientist, I am not able to make ridiculous claims, such as, *"mankind is better off without a religious belief system".*

I am proud to say to you that, I am a scientist and I accept all religions to be biologically true and equal. My pursuit of understanding the human mind has taught me universal tolerance. I shall quote a few lines which I remember to have repeated time and again in the past:

"As the different streams having their sources in different paths which men take through different

tendencies, various though they appear, crooked or straight, all lead to Thee."

If the history of mankind has shown anything to the world, it is that piety, purity and charity are not the exclusive possessions of any church, temple, mosque or synagogue in the world, and that every system has produced men and women of the most glorious character. In the face of this evidence, if anybody dreams of the exclusive survival of his or her own religious or non-religious belief system and the destruction of the others, I pity that person from the bottom of my heart. I want to point out to those heinous individuals that we are at the dawn of a new world, where upon the banner of every religion will soon be written in spite of resistance: "Help and not fight," "Assimilation and not Destruction," "Harmony and Peace and not Dissension."

In the present book, I don't make any pretence of knowing about the existence of a Supreme Entity, neither do I make any attempt to create any friction among religions. If anything, I have spared myself no pains in my endeavor to smoothen the ongoing friction among all religions of the world. May the book give you a few hours of suggestive thoughts and direct you in the path of sweet general harmony.

And if anyone hopes that the present book will end up being an aid for the triumph of any one religion and the destruction of all others, to that person I say, *"my dear friend, yours is neurologically an impossible hope"*. The same goes for atheism.

Based on one's own needs, his or her brain develops various belief systems, be it religious, non-religious or spiritual. When a seed is put in the ground, and earth, air and water are placed around it, does the seed become the earth, or the air, or the water? No. It becomes a plant. It develops according to the law of its own growth, assimilates the air, the earth, and the water, converts them into plant substance, and grows into a plant.

Similar is the case with belief system. The Christian is not to become a Muslim or a Hindu, nor a Muslim or a Jew to become a Christian or a Hindu. In the name of harmony, each must assimilate the spirit of the others yet preserve his or her individuality and grow according to his or her own law of growth.

The only way a planet can move forward in the path of harmony and development, is through universal tolerance. And whenever general harmony is in danger I shall rise time and time again from the ashes of my own past.

*   *   *

PREFACE

Mankind is a vessel filled with vivid and fascinating features. All of these features have developed through millions of years of evolution. And all of them serve only one purpose, - *survival of the species*. In the face of survival there is nothing right or wrong. In fact, our ability to distinguish the right from the wrong is just another feature of our brain's frontal lobes. Such extraordinary cognitive ability has been endowed on us by Mother Nature, so that we can choose the path that is relatively most suitable for our survival.

Among the countless evolutionary traits that we possess, the most crucial ones are only a few — *curiosity*, *sexuality* and *spirituality*. Without these three, we would have gotten extinct long ago just like our hominin ancestors.

Curiosity developed in the hominin brain while living in the wild, being overwhelmed with fear and ignorance for millions of years. It developed as a counter mechanism to find answers to the questions we could not even conceive back then.

Sexuality allows us to copulate and pass on our genetic traits through time. It makes us immortal in a manner of speaking.

Last but definitely not the least, **spirituality is the evolutionary trait that allows us to see *the big picture*,** regardless of whether there is a picture. The purpose of this trait is not to see the true nature of the universe. Rather it is a kind of random calibration mechanism of the brain, through which often complex problems of life can be solved.

Curiosity and sexuality are easy to describe and understand. Spirituality on the other hand involves all kinds of mysterious and bizarre characteristics, some of which may seem to be absolutely non-sense to the untrained eye. Precisely that is the point. The very nature of spirituality is personal and individualistic. There is no correct universal definition of spirituality. Because the very foundation of it is deep-rooted into the individual's unique right hemispheric circuitry of the brain. In short, spirituality is personal.

The meanings of spirituality are vividly different for different humans. Different humans interpret spirituality in different manners. For example, some people experience spirituality in their creativity. To some, spirituality is in the art or music they create. Whereas, to some others it can be science itself. In all these cases, those humans may or may not be typically religious.

The human brain has an innate urge to make sense even out of non-sense. For this purpose the brain always tends to construct parameters. These structural parameters enable us to avoid further strenuous intellectual efforts. In time, through human interpretation these parameters slowly transform into the form of conventional *religions*.

And if history has shown anything, it is that the first person to construct such parameters always ends up becoming a venerated figure in his or her specific line of convention. Later on, the individuals in that line of religious convention slowly start to correlate their own existential identity with the identity of the patriarch, i.e. the person (mostly man, not woman) who first constructed the parameters of their religious convention. Naturally it becomes imperative that they protect and promote the identity of their venerated figure or patriarch at all costs. And one simple way to do so, is to deny the authenticity of all other religious conventions and their patriarchs. For this specific purpose, some humans come up with various innovative psychological tricks, such as referring to their patriarch as *"the last prophet"*. To the human mind, such phrase itself demolishes all possibility of any further arrival of prophets or patriarchs. It is all about self-interest and self-maintenance. It is the brain's way of sustaining one's identity.

Any threat to the religious convention or to the patriarch feels like a threat to one's identity. And a threat to the identity is the psychological equivalent of a physical threat to survival.

Henceforth, the individuals in a line of religious convention sometimes construct further parameters and principles to venerate the life of the patriarch, for example the *Hadith*. And these principles and parameters often tend to define how a person in that religious convention should live life.

Eventually all these parameters, be it constructed by a religious patriarch or by his followers, get compiled into various books, which we humans term as *scriptures*. To the layman (uneducated as well as highly educated intellectuals), these scriptures act as Instruction Manuals of life, thus relieving the person from all intellectual efforts of finding answers to the big questions.

When we go deeper into the origin of those Instruction Manuals or the *scriptures*, we discover something extraordinary. At the root of all the scriptures there is just another plain ordinary human mind. All the messages in the scriptures were constructed by the human mind in a transcendental state of consciousness. Hence, they have no connection with any kind of Supreme Entity whatsoever. There is no such thing as *God given scripture, the messenger of God* or *the Last Prophet*. Regardless of the geography, all the scriptures of mankind history were created by the humans for the humans, be it, the Bible, the Vedas, the Quran or any other.

Every time, an ordinary human being goes into a transcendental state of consciousness through whatever means (prayer, meditation, brain lesion, geomagnetic disturbance, psychedelics), he or she experiences such an extraordinary and inexplicable bliss that it seems something heavenly. In this state of mind, all the previous confusions and worries get resolved through the person's subconscious mind in the form of sacred messages from God. Naturally, when the person comes out of that altered state of consciousness, his or her entire personality changes with the invincible belief that he or she has accessed the mind of God. The person starts believing that his or her experience is worth pursuing by the whole of humanity. Therefore, he or she starts preaching the sacred messages to the masses given to him or her by the simple three pound lump of jelly.

ONE

THE PROPHETS

Our universe is a human universe, because our brain constructs our understanding of the universe with our limited range of perception. As biological creatures we shall never be able to perceive the true nature of the universe. Every model of the universe, we construct with our human brain is just another perceptual illusion with only a few elements of absolute truth in it.

Just for a second, think, how mysteriously vast the universe is! And we the humans exist only in a tiny fraction of that vastness. We can realize how insignificant we are if we compare ourselves with the vastness of the universe. Our universe is everything that is out there. The spongy grey matter on our head, called the brain has access to only a microscopic percentage of that unfathomable everything. We childishly boast our greatness as a so-called advanced species while we only see a very small strip of what's really going on in the universe.

The human brain receives tiny bits of information from the surroundings and fills the gaps with conjectures, beliefs and fantasies. All the human brains on earth construct their own idea of the universe and how things in it work. The brain does this through developing mechanisms of beliefs, be it religious, non-religious, political or others. All of us humans have our own unique belief systems. And as humans, it is excruciatingly hard to step outside that circle of belief, unless one is a scientist studying the beliefs.

Let's conduct a thought experiment that will illustrate the variance in belief systems in the simplest and most intelligible form.

Once upon a time, there was a frog that lived in a well. It had lived there for a long time. It was born there and brought up there, and yet was a little, small frog. One day another frog that lived in the sea came and fell into the well.

"Where are you from?"

"I am from the sea."

"The sea! How big is that? Is it as big as my well?" and it took a leap from one side of the well to the other.

"My friend," said the frog of the sea, *"how do you compare the sea with your little well?"*

Then the frog took another leap and asked, *"Is your sea so big?"*

"What nonsense you speak, to compare the sea with your well!"

"Well, then," said the frog of the well, *"nothing can be bigger than my well; there can be nothing bigger than this; this fellow is a liar, so turn him out."*

This has been the difficulty with religious beliefs all through ages. A Christian sits in his or her well and thinks that the whole world is his or her well. The Jew sits in his or her little well and thinks that it is the whole world. A Muslim sits cooped up in his or her tiny well and believes it to be the whole universe. The same goes for a Hindu and all others. Also, atheists are no different from all those orthodox believers.

All the tiny wells of religious beliefs were constructed by humans, upon the spiritual experiences of a few ordinary humans such as Abraham, Moses, Vyasa, Valmiki, Jesus, Mohammed, Nanak, Siddhartha Gautama, Mahavira, Joseph Smith, and so on. Naturally all these individuals became historical figures of utmost veneration in the wells that their followers created.

In these little circles of beliefs, humans remain cooped up and tend to foster their personal ideas of the universe through the teachings of those religious legends. As a psychological protective mechanism, we even come up with unique titles to impose on those patriarchs. Jesus is widely venerated as *the son of God*, Siddhartha Gautama as *Buddha* or *The Enlightened One*, Nanak as the *Guru* and Mohammed as *The Last Prophet.*

It is neither stupid nor idiotic. Rather it is how the human brain works. If there is anyone to blame, then it is *evolution*. As humans we always tend to simplify things through a reductionist approach. No matter how sophisticated and intelligent a person poses to be, his or her brain still works in a reductionist approach to everything. The brain does this, because it suits our needs of understanding, no matter what the reality is. For example, if there is a God, (which I don't know) then, aren't we all his or her children? On the other hand, anybody who experiences the awakening of the mind through an altered state of consciousness is an Enlightened One, or a Buddha.

Likewise, there is no such thing as the last prophet. The English word prophet comes from the Greek word *profétés* meaning advocate or speaker. As per the definition of the term, any human being who tends to speak the language of God is hailed as a prophet. So, every single person throughout history, who has ever had a transcendental experience is technically a prophet.

Historically, we know a few individuals who can be hailed as the early prophets of mankind, such as Noah, Abraham, Vyasa, Valmiki, Moses and so on. But putting aside the theological mumbo-jumbo, there is no *last* in the line of prophets. Because till this day in our clinical studies we come across many individuals who report having an encounter with God or angels. And if modern Neuroscience has demonstrated anything, it is that, terms like *prophet, enlightened one, rishi, Guru* etc. are not exclusive possessions of any mosque, temple, church or synagogue.

It is all about experiencing the ultimate transcendental state of consciousness. This state of the human mind, amplifies a person's own desires, instincts, fantasies and urges. It has no supernatural element involved. It only feels supernatural, because the brain makes you feel so. In various ancient cultures throughout the world, this heightened state of the human mind has been worshipped with utmost reverence. This is what all the patriarchs of religions experienced. This is what they explained as their blissful encounter with God or the Universal Consciousness. The same state of mind has been experienced by different individuals in different geographical regions at different times. It is what Jesus experienced through his prayers and devotion. Abraham, Moses and Mohammed felt it through prayers. Buddha and all the Indian sages such as Vyasa (author of the Mahabharata), Valmiki (author of Ramayana), Patanjali (author of Yogasutra) attained that state through intense meditation. None of them is superior to another. All of them wanted answers to the mind-boggling questions of life so badly, that their own subconscious mind created the answers and delivered them to their conscious mind when they were immersed in a state of transcendence.

17

When you start to enter the realm of transcendence, first comes the loss of all physical awareness, then you cease to identify yourself with the mind, but a shade of ego is left. Then this little shade of ego or "I" also disappears, as the parietal lobes of your brain shut down. At that point whatever remains is the purest form of bliss you can ever experience. Once you attain this state there is no intellectual difference between you and all those religious leaders.

Any human being who goes into the hyperspace of transcendence, comes out as a reformed person (never as an absolute embodiment of goodness and kindness, but with both good and bad elements), because in that state, when your parietal lobes shut down, you lose all touch with reality. Literally speaking, everything that keeps you inside a tiny circle of perception, suddenly disappears. Everything that is conventional and that separates you from the rest of the world in the name of *uniqueness* suddenly vanishes. You experience the true union with the universe. As if your perception knows no bounds.

Let me tell you the story of Nanak, the founder of Sikhism, to elucidate the *oneness* of transcendence. Nanak used to take bath in the river every day before sunrise. But one day he mysteriously disappeared with his clothes still lying on the river bank. Three days later he emerged from the water uttering *"there is neither Hindu nor Muslim"*.

Various fascinating psychological elements are involved in the transcendental state of human consciousness. One may lose the ability to distinguish one's self from the rest of the world in transcendence, but still it is the human brain that constructs that state of mind. Hence, even in that altered state of consciousness one is not totally devoid of one's beliefs, conjectures, ideas and fantasies. In fact, these ideas fill up the transcendental experience with all kinds of fanatic stories that happen to be unique, based on the person's inner urges and drives. And unless you are a person who has spent his or her entire life studying these experiences, you can never perceive the transcendental visions to be the creation of your own brain. To you it would seem *more real than real*. And this has been true for all the religious figures of human history.

For example, Nanak's divine encounter with God triggered a whole new religious movement called *Sikhism*. While talking about his religious experience he described that he was taken to the court of God and given a cup of amrit (divine nectar) to drink. Then God commanded him *"I am with you, Go and repeat My Name, and teach others to do the same"*. He was so much filled with divine bliss due to the flood of dopamine in his body, that he composed the first foundational verse, - the Mul Mantar (Root Mantra) for the central scripture of Sikhism, Guru Granth Sahib, which is considered to be the eleventh Guru in that religious tradition. The entire Sikh faith is based upon Mul Mantar:

"Ik Onkar, Sat Nam, Karta Purakh, Nir Bhau, Nir Vair, Akal Murat, Ajuni, Saibhang, Gur Prasaad"

Translation:

"There is only one God,

Eternal truth is His name,

He is the creator,

He is without fear,

He is without hate,

He is without form,

Beyond birth and death,

He is the enlightener,

He can be reached through the mercy and grace of the true Guru (teacher)."

Quite similarly, in the Islamic circle of belief, Mohammed is the historic figure of highest importance. Due to the innate psychological urges, many of the muslims consider him to be *the last prophet sent by God.* The Islamic faith of Quran is based on the spiritual experiences of Mohammed. In the Islamic circle of belief, the individualistic human brains make people honestly think that after Adam, Noah, Abraham, Moses and Jesus, God finally sent the last prophet and messenger Mohammed to deliver the spiritual teachings. He is perceived to have restored the unaltered origin of monotheistic faith of Adam, Abraham, Moses and Jesus in Islam.

Mohammed was born approximately in 570 CE in the Arabian city of Mecca. Being orphaned at an early age, he was raised under the care of his paternal uncle Abu Talib. He primarily worked as a merchant. Occasionally he would retreat to a cave in the mountain Jabal al-Nour, for several nights of solitude and prayer. It was there in a cave called Hira, at the age of forty during his prayers, he had his first transcendental state of consciousness in which he encountered the Archangel Gabriel.

21

As per the perceptual story created by his own human brain during the transcendence, Archangel Gabriel commanded him to recite the first verse for Quran. It was later recorded in the Quran as the first Surah revealed to Mohammed - the Surah Al-Alaq (chapter 96).

QURAN : SURAH 96 (AL-ALAQ), AYAT 1-5

Recite in the name of your Lord who created –

Man from a clinging substance.

Recite, and your Lord is the most Generous –

Who taught by the pen –

Taught man that which he knew not.

Three years after this event Mohammed started preaching these revelations of Quran to people, declaring the oneness of God. Thus he created a whole new circle of religious belief system. A similar event took place in the life of Joan of Arc at a different time under different circumstances.

Transcendental experiences can be triggered by various physical stimuli. Mohammed had his experience through prayers, whereas Joan of Arc's experience was caused by brain lesion. She suffered from temporal lobe epilepsy. During her epileptic fits, she had hallucinations of various divine entities, just like Mohammed had during his prayers.

Joan of Arc was born in Domremy, 1412. Her father has variously been described as a ploughman, keeper of a cattle pound, and a low-ranking official. Her early childhood, in the village where she was born, has been described as harsh and she received no formal teaching though her mother taught her basic household skills. She also tended her father's cattle, though legend has it that she was a shepherdess. She was about 13 when she first heard voices. The description of this first experience is quoted in Smith's Joan of Arc:

"She had a voice from God to help her to know what to do. And on this first occasion she was very much afraid ... She heard the voice upon the right side and rarely heard it without accompanying brightness ... after she heard this voice upon three occasions, she understood that it was the voice of an angel".

Later, she went on claiming that she had visions of the Archangel Michael, Saint Margaret and Saint

Catherine instructing her to support Charles VII and recover France from English domination, just like Mohammed did with spreading the Quran.

Mohammed was a plain ordinary human just like you and me, and so was Abraham, Moses, Manu, Vyasa, Jesus, Buddha, Joan of Arc, Dostoyevsky and all the people who had the divine experience of transcendence.

The transcendental state of consciousness is a goldmine of philosophical and metaphysical information suitable for the specific person who goes into that state. In this altered state, the human brain goes into hyper-drive, and starts firing rapid electrochemical signals through the synapses involving all of a person's beliefs, conjectures, ideas, perceptions, confusions, urges and fantasies. And as the end product of such complex and stormy neural activity, the person apparently receives information of great cosmic significance. One who experiences this, becomes absolutely convinced that he or she has accessed the cosmic consciousness of the universe. And in an attempt to explain the experience in his or her native language, the person ends up creating conflicts between his or her own perception of the universe and all other religious belief systems.

Consciousness in a state of transcendence, is not bound by our familiar laws of nature, despite being within the nature. Being completely disconnected from our usual reality, this altered state of consciousness is all about feeling absolutely free. But the moment, the human mind shifts from the transcendental state back to the usual waking state, it suddenly finds itself tied to the laws of various natural processes.

And when it comes to expressing the experience in plain human words, things get extremely bumpy. Language is a very limited process. And every single word a person uses to depict his or her divine encounter, has significant impact over the followers' thinking process. Those words uttered by the lips of an *enlightened being* slowly fill up the void inside the pupils' mind, just like hot tea coming out of the nozzle of a kettle fills up an empty vessel. Over time, those powerful words of the *prophet* become indelibly engraved into the conscious, sub-conscious and un-conscious mind of the followers.

In time, to those hardcore followers, the words uttered by their own religious leader become the only acceptable expression of the Supreme Divinity. Ergo, their tiny little human brains often perceive all other divine encounters of mankind history to be false. Hence, the humans come up with apparently mortifying and racist terms, such as *"kafir"* (any non-Muslim), that don't have any true significance whatsoever.

The human brains in some religious circles often construct laws to prevent any kind of event that may endanger the identity of their religious patriarch, because any potential threat to, the identity of the patriarch, the God that he talked about, or the entire belief system, is a threat to the very existence of their religious identity. Therefore, ordinary human words involving nothing but heinous self-interest crawl onto the pages of the scriptures and hide underneath the blinding shine of the sacred texts.

THE HEBREW BIBLE, EXODUS 34: 11-17

"(God said) "Be sure to observe what I am commanding you this day: behold, I am going to drive out the Amorite before you, and the Canaanite, the Hittite, the Perizzite, the Hivite and the Jebusite. Watch yourself that you make no covenant with the inhabitants of the land into which you are going, or it will become a snare in your midst. But rather, you are to tear down their altars and

*smash their sacred pillars and cut down their Asherim --
for you shall not worship any other god, for the LORD,
whose name is Jealous, is a jealous God-- otherwise you
might make a covenant with the inhabitants of the land
and they would play the harlot with their gods and
sacrifice to their gods, and someone might invite you to
eat of his sacrifice, and you might take some of his
daughters for your sons, and his daughters might play
the harlot with their gods and cause your sons also to
play the harlot with their gods. You shall make for
yourself no molten gods.""*

SAHIH MUSLIM 1:30

*"It is reported on the authority of Abu Huraira that the
Messenger of Allah said: I have been commanded to fight
against people so long as they do not declare that there is
no god but Allah, and he who professed it was
guaranteed the protection of his property and life on my
behalf except for the right affairs rest with Allah."*

SAHIH MUSLIM 1:32

*"It is narrated on the authority of Jabir that the
Messenger of Allah said: I have been commanded that I
should fight against people till they declare that there is*

no god but Allah, and when they profess it that there is no god but Allah, their blood and riches are guaranteed protection on my behalf except where it is justified by law, and their affairs rest with Allah, and then he (the Holy Prophet) recited (this verse of the Holy Qur'an):" Thou art not over them a warden""

SAHIH MUSLIM 1:33

"It has been narrated on the authority of Abdullah b. 'Umar that the Messenger of Allah said: I have been commanded to fight against people till they testify that there is no god but Allah, that Mohammed is the messenger of Allah, and they establish prayer, and pay Zakat and if they do it, their blood and property are guaranteed protection on my behalf except when justified by law, and their affairs rest with Allah."

SAHIH MUSLIM 1:34

"It Is narrated on the authority of Abu Malik: I heard the Messenger of Allah (may peace be upon him) say: He who professed that there is no god but Allah and made a denial of everything which the people worship beside Allah, his property and blood became inviolable, an their affairs rest with Allah."

On planet earth, survival is everything for a biological creature such as humans. And constructing Instruction Manuals with apparently sacred laws is just an element of the brain's complex self-maintenance mechanism to ensure survival. It has nothing to do with any kind of Supreme Creator, if there is any.

The mechanism of religion is about being content with what we have. Through the true spirit of religiosity and spirituality, one can truly learn to train the mind and direct it towards a better future. It is you, who decides how to move forward in the path of true enlightenment, not some ancient literature. Open your eyes to new possibilities and assimilate the elements of kindness from all the religious characters throughout the world. And simply ignore the dark elements of the scriptures. After all, it is not about knowing the prophets, their lifestyles, the scriptures or even God for that matter. It is all about knowing yourself while being under your own skin. Religion is about discovering the prophet within you. It is about evolving God out of humans. One who sees the child, sees the Supreme Father or Mother, if there is any.

After all, it is the transcendental experience of the human brain that shines upon the followers of

Christ, Jehovah, Krishna, Allah, Buddha and all other religious leaders, like one radiant and glorious sun.

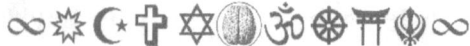

TWO

MISOGYNY

Most religious scriptures are all about three apparently glorifying elements, - *man, man* and *man*. And it doesn't take a neuroscientist to realize the reason behind it. Almost all the scriptures are filled with the primitive element of misogyny, because all of them were authored by men. That's why it is more common for us to say *religious patriarchs*, than *religious matriarchs*. A Jewish fundamentalist would say, it is God's will for a man to be in control of everything on earth, not a woman, because God created woman out of a rib from the man's body, not the other way around. A Muslim extremist would say, man is the master of a woman, since a woman is deficient in intellect. And all these typical conservative beliefs are derived from the textual interpretation of the scriptures.

But, as I have elaborated in the last chapter, those so-called Instruction Manuals have nothing to do with any kind of Supreme Creator whatsoever. Every single word of the scriptures, be it good or bad, was born inside the human brain, from the little wisps of protoplasm. Not a single word came from any celestial source.

All those sacred words were constructed by humans in order to suit the personal needs of the humans. Almost all of those compilers who despite their childish stupidity, succeeded in pretending to be brilliant scholars, were ordinary men. One cannot expect them to waste any opportunity of placing their own darkest desires and heinous urges on the sacred pages of the religious texts as the command from God. That is what any ordinary human would do. However, I am not so sure about a true human being.

Now let me ask you a simple question.

What's the point of having so many sophisticated scriptures and highly venerated prophets if they do not possess the simple ability to teach us plain ordinary everyday kindness for both man and woman?

Neglect of women is the major cause for the society's downfall. The man and the woman are the two wheels of the society. If either one becomes defective, the society cannot make progress. There will be hope for the well-being of the entire world only if the humans, men and women alike, stop deeming the women as some kind of inferior creatures. We, the responsible citizens of the world can build a better future for the entire species, only by improving the condition of the women. And here I am not advocating for feminism. I am simply elucidating one rudimentary element of the path towards Universal Tolerance.

It is very unfortunate that only a few of the world religions put woman in the same pedestal as the man. And major religious scriptures of the world such as Judaism, Hinduism and Islam are absolutely opposite and downright monstrous on this matter, since they are filled with unambiguous misogynic verses made up by their compilers. To talk about gender-equality in religion, the most important religious, or rather spiritual figure that comes to my mind is Siddhartha Gautama. In the gracious eyes of this great awakened human being, so named *Buddha*, gender does not exist in religion.

To him, gender was of complete irrelevance to religion.

The glorious character of Buddha shines like a radiant sun with nothing but deepest kindness and respect for the women. He was a simple person who lived a simple monastic life and was totally incapable of being a heartless misogynist. The very core of Buddha's philosophical teachings is based upon boundless kindness and compassion towards all beings.

In many cases Buddha advised the householders about the roles and status of the two genders which must have stood out in his culture for the reciprocity and mutual respect he recommended. Take the following verse from one of the Buddhist scriptures, *Digha Nikaya* for instance, where he described the respective duties of husbands and wives.

DIGHA NIKAYA 31

"In five ways should a wife as Western quarter, be ministered to by her husband: by respect, by courtesy, by faithfulness, by handing over authority to her, by providing her with ornaments. In these five ways does

the wife minister to by her husband as the Western quarter, love him: her duties are well-performed by hospitality to kin of both, by faithfulness, by watching over the goods he brings and by skill and industry in discharging all business."

In this context, I can't help but bring up a few verses from the Holy Quran and the Hadith, provided by Mohammed as advice to the householders.

SAHIH AL-BUKHARI 7:62:121

"Narrated Abu Huraira:

The Prophet said, "If a man Invites his wife to sleep with him and she refuses to come to him, then the angels send their curses on her till morning.""

SAHIH AL-BUKHARI 7:62:122

"Narrated Abu Huraira:

The Prophet said, "If a woman spends the night deserting her husband's bed (does not sleep with him), then the angels send their curses on her till she comes back (to her husband).""

QURAN, SURAH 4 (AN-NISA), AYAT 34

"Men are in charge of women by [right of] what Allah has given one over the other and what they spend [for maintenance] from their wealth. So righteous women are devoutly obedient, guarding in [the husband's] absence what Allah would have them guard. But those [wives] from whom you fear arrogance - [first] advise them; [then if they persist], forsake them in bed; and [finally], strike them. But if they obey you [once more], seek no means against them. Indeed, Allah is ever Exalted and Grand."

I shall not go into the theological debate about how much misogynic Mohammed was, or whether he was a man of peace or violence. It is not up to my standards. Even a toddler can tell that, those words were simply the product of Mohammed's own mind playing tricks in order to satisfy his deepest and darkest desires. So, his desires took the form of sacred command, when he was having what we may call apparently the divine experience of a lifetime. It is not really his fault, since during an altered state of consciousness such as in transcendence, it is almost impossible for any human being to tell the right from the wrong, and the good from the bad.

One's personality influences those sacred messages received in transcendence to a large extent. After all Mohammed, Buddha and all others were simple humans. But what they learnt in their altered state of mind is distinctively different from each other, due to their own unique personalities, urges, beliefs and fantasies. A human being has both good and bad elements in his or her personality. Biologically we are predisposed to be either bad or good at times, based on the need of the circumstances. It is all inside our head.

The limbic system of our brain, holds all our deepest and darkest secrets. It is the house of our kinkiest desires. This wild feature of the human mind is what we may call the *"Id"*, in psychoanalytic terms. Id is the wild beast within each one of us. This beast is reckless and beyond all social norms. This drives our innate evolutionary instinct of surviving even in the harshest climate. Because, nature doesn't give a damn about any living creature. She exists and will keep on existing, with or without us. It is us who needs nature, not the other way around. For this exact reason, in primitive days humans had to fight hard in the wild environment of Mother Nature, against all calamities. And this age-old fight for survival has molded the human brain only as per the parameters of survival. Whatever we are today, is a product of millions of years long constant battle for survival. Nature selects biological features as well as destroys them when deemed unnecessary.

In the harsh climate of the wild, the limbic system may have been our best bet for survival, but as we started to understand our uniqueness, things began to change. We understood the power of community. We realized that, together we could fight even the fiercest predators, but alone we'd be simply their prey. Naturally, Mother Nature put selective pressure on the human brain to evolve significantly, making it go through complex neural reorganization while increasing in size to a great extent then decreasing a little.

Human brain is the organ of unparalleled importance. It is the mother organ that drives every single mechanism in the body. It is the organ with which you see, perceive, observe, think, reminiscence, pleasure and even carry out feats of spiritual or mystical significance such as sensing an incomprehensible supernatural air in the world, giving it a comprehensible form through language and ultimately creating the scriptures while depicting the source of those scriptures in some Supreme Invisible Entity.

Essentially, the evolution of the human brain has been one of the most significant events in the evolution of hominin life. It has been a 6 million years long mosaic process of size increase laced with episodes of reorganization of the cerebral cortex. And the human brain is the most intricate, complicated and impressive organ ever to have evolved on planet earth. The most obvious evolutionary change during human evolution has been the increase in size and complexity of the human brain.

The study of the evolution of human brain is an intriguing domain of scientific exploration. It's like talking to the ghosts of our extinct ancestors through their fossil remnants. This is what we call "Paleoneurology". Paleoneurology allows us to look into the details of brain structure of extinct species through close observation of endocasts (endocranial casts). It is a subfield of paleoanthropology. Paleoanthropological studies show us that social organization was imperative for the early hominins to survive in the harsh environment. So, natural selection forced the brain to develop primitive social interaction through gesture and mimicry. Such selective natural pressure gave our early Australopithecine ancestors cortical capacity for social coordination. Also, any negative emotional outbreak would attract the attention of the predators, so our early ancestors had to develop emotional control. As the early humans gained control over their emotions by the augmentation of the prefrontal cortex, their social communities became more stable. This is exactly when our ancestors took the first step towards being civilized humans from wild beasts.

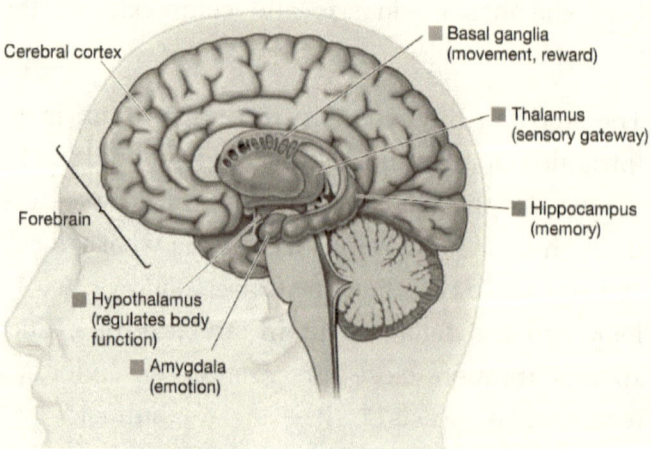

Figure 2.1 The Limbic System surrounded by the layers of Cerebral Cortex

This was all because of the augmentation in the frontal lobes, especially the prefrontal cortex. The prefrontal cortex (PFC) is the brain region that keeps the primitive beast within the limbic system from lashing out. From a psychoanalytic perspective, we can say that PFC houses the *Ego,* that keeps all our emotions in check. In usual waking state of consciousness, the PFC decides which emotion at what intensity should be expressed depending on the situation. A lot of factors involves in this mechanism, including social norms and the common distinction between right and wrong.

But the moment a human brain goes into the state of transcendental consciousness, all those factors vanish. In this state the human mind is biologically incapable of perceiving any kind of social norms and even the simplest distinction between right and wrong. At this state a person's true personality finds its way out. Hence, a nomadic personality gives rise to a scripture that is filled with nomadic instincts of imperialism. A misogynic personality constructs a scripture filled with depictions of inferiority of women. Whereas decent personalities develop philosophical teachings of kindness, love and compassion, that may have only a few dark elements, since all of them are humans.

The point is, as long as the concept of divinity resides within the human heart as a sense of spirituality, peace and harmony prevail. But the moment, humans try to express this sensation in words, it is destined to have substantial impact over the society. Now it is up to you, whether you choose the good elements of those philosophical teachings, or simply gobble everything that is in there, good and bad elements alike, thus halting the progress of the society. In order to elucidate on this matter, I shall bring up some unambiguous verses from the Islamic, Jewish and Hindu scriptures. I shall not comment on whether they consist of the religious elements of true humanity or plain primitive madness. Any modern human being with a lucid mind can do so.

SUNAN ABU DAWUD 11:2142, 2155

"Mohammed said: A man will not be asked as to why he beat his wife.

Mohammed said: If one of you marries a woman or buys a slave, he should say: "O Allah, I ask You for the good

*in her, and in the disposition You have given her; I take
refuge in You from the evil in her, and in the disposition
You have given her." When he buys a camel, he should
take hold of the top of its hump and say the same kind of
thing."*

QURAN, SURAH 2 (AL-BAQARAH), AYAT 282

*"O you who have believed, when you contract a debt for
a specified term, write it down. And let a scribe write [it]
between you in justice. Let no scribe refuse to write as
Allah has taught him. So let him write and let the one
who has the obligation dictate. And let him fear Allah ,
his Lord, and not leave anything out of it. But if the one
who has the obligation is of limited understanding or
weak or unable to dictate himself, then let his guardian
dictate in justice. And bring to witness two witnesses
from among your men. And if there are not two men
[available], then a man and two women from those
whom you accept as witnesses - so that if one of the
women errs, then the other can remind her. And let not
the witnesses refuse when they are called upon. And do
not be [too] weary to write it, whether it is small or
large, for its [specified] term. That is more just in the
sight of Allah and stronger as evidence and more likely
to prevent doubt between you, except when it is an*

immediate transaction which you conduct among yourselves. For [then] there is no blame upon you if you do not write it. And take witnesses when you conclude a contract. Let no scribe be harmed or any witness. For if you do so, indeed, it is [grave] disobedience in you. And fear Allah . And Allah teaches you. And Allah is Knowing of all things."

SAHIH AL-BUKHARI 1:6:301

"While on his way to pray, Mohammed passed a group of women and he said, "Ladies, give to charities and donate money to the unfortunate, because I have witnessed that most of the people in Hell are women.

They asked, "Why is that?"

He answered, "You swear too much, and you show no gratitude to your husbands. I have never come across anyone more lacking in intelligence, or ignorant of their religion than women. A careful and intelligent man could be misled by many of you."

They responded, "What exactly are we lacking in intelligence or faith?"

Mohammed said, "Is it not true that the testimony of one man is the equal to the testimony of two women?"

After they affirmed that this was true, Mohammed said, "That illustrates that women are lacking in intelligence. Is it not also true that women may not pray nor fast during their menstrual cycle?" They said that this was also true.

Mohammed then said, "That illustrates that women are lacking in their religion.""

SAHIH MUSLIM 17:4206

"There came to Mohammed a woman who said: Allah's Messenger, I have committed adultery, [...] When she was delivered she came with the child (wrapped) in a rag and said: Here is the child whom I have given birth to. He said: Go away and suckle him until you wean him. When she had weaned him, she came to him with the child who was holding a piece of bread in his hand. She said: Allah's Apostle, here is he as I have weaned him and he eats food. He entrusted the child to one of the Muslims and then pronounced punishment. And she was put in a ditch up to her chest and he commanded people and they stoned her."

SAHIH AL-BUKHARI 7:62:124

"Narrated Usama:

The Prophet said, "I stood at the gate of Paradise and saw that the majority of the people who entered it were the poor, while the wealthy were stopped at the gate (for the accounts). But the companions of the Fire were ordered to be taken to the Fire. Then I stood at the gate of the Fire and saw that the majority of those who entered it were women.""

Sahih Muslim 3:0684

"Abu Musa then said, "When is a bath obligatory?" Aisha responded, "You have asked the right person. Mohammed has said that a bath is obligatory when a man is en- compassed by a woman and their circumcised genitalia touch.""

The Hebrew Bible, Genesis 2:21-23

"And the LORD God caused a deep sleep to fall upon Adam, and he slept: and he took one of his ribs, and closed up the flesh instead thereof And the rib, which the LORD God had taken from man, made he a woman, and brought her unto the man. And Adam said, This is now

bone of my bones, and flesh of my flesh: she shall be called Woman, because she was taken out of Man."

THE HEBREW BIBLE, EXODUS 20:17

"Thou shalt not covet thy neighbour's house, thou shalt not covet thy neighbour's wife, nor his manservant, nor his maidservant, nor his ox, nor his ass, nor any thing that is thy neighbour's."

THE HEBREW BIBLE, EXODUS 21:7-11

"And if a man sell his daughter to be a maidservant, she shall not go out as the menservants do. If she please not her master, who hath betrothed her to himself, then shall he let her be redeemed: to sell her unto a strange nation he shall have no power, seeing he hath dealt deceitfully with her. And if he have betrothed her unto his son, he shall deal with her after the manner of daughters. If he take him another wife; her food, her raiment, and her duty of marriage, shall he not diminish. And if he do not these three unto her, then shall she go out free without money."

THE HEBREW BIBLE, LEVITICUS 21:7,9

"They shall not take a wife that is a whore, or profane; neither shall they take a woman put away from her husband: for he is holy unto his God.

And the daughter of any priest, if she profane herself by playing the whore, she profaneth her father: she shall be burnt with fire."

THE HEBREW BIBLE, NUMBERS 1:2

"Take ye the sum of all the congregation of the children of Israel, after their families, by the house of their fathers, with the number of their names, every male by their polls."

THE HEBREW BIBLE, NUMBERS 30:1-16

"And Moses spake unto the heads of the tribes concerning the children of Israel, saying, This is the thing which the LORD hath commanded. If a man vow a vow unto the LORD, or swear an oath to bind his soul with a bond; he shall not break his word, he shall do according to all that proceedeth out of his mouth. If a woman also vow a vow unto the LORD, and bind herself by a bond, being in her father's house in her youth; And her father hear her vow, and her bond wherewith she

hath bound her soul, and her father shall hold his peace at her: then all her vows shall stand, and every bond wherewith she hath bound her soul shall stand. But if her father disallow her in the day that he heareth; not any of her vows, or of her bonds wherewith she hath bound her soul, shall stand: and the LORD shall forgive her, because her father disallowed her. And if she had at all an husband, when she vowed, or uttered ought out of her lips, wherewith she bound her soul; And her husband heard it, and held his peace at her in the day that he heard it: then her vows shall stand, and her bonds wherewith she bound her soul shall stand. But if her husband disallowed her on the day that he heard it; then he shall make her vow which she vowed, and that which she uttered with her lips, wherewith she bound her soul, of none effect: and the LORD shall forgive her. But every vow of a widow, and of her that is divorced, wherewith they have bound their souls, shall stand against her. And if she vowed in her husband's house, or bound her soul by a bond with an oath; And her husband heard it, and held his peace at her, and disallowed her not: then all her vows shall stand, and every bond wherewith she bound her soul shall stand. But if her husband hath utterly made them void on the day he heard them; then whatsoever proceeded out of her lips concerning her vows, or concerning the bond of her soul, shall not stand: her husband hath made them void; and the LORD shall forgive her. Every vow, and every binding oath to afflict the soul, her husband may establish it, or her husband may make it void. But if her husband altogether hold his peace at her from day to day; then he establisheth all her vows, or all her bonds, which are upon her: he confirmeth them, because he held his peace at her in the day that he heard them. But if he shall any ways make them void after that he hath heard them; then he shall bear her iniquity. These are the statutes, which the LORD commanded Moses, between a man and his wife, between the father and his daughter, being yet in her youth in her father's house."

The Hebrew Bible, Deuteronomy 24:1

"When a man hath taken a wife, and married her, and it come to pass that she find no favour in his eyes, because he hath found some uncleanness in her: then let him write her a bill of divorcement, and give it in her hand, and send her out of his house."

Manusmriti VIII:1

"If a wife, proud of the greatness of her relatives or (her own) excellence, violates the duty which she owes to her

*lord, the king shall cause her to be devoured by dogs in a
place frequented by many."*

MANUSMRITI IX:1-3

*"I will now propound the eternal laws for a husband and
his wife who keep to the path of duty, whether they be
united or separated.*

*Day and night woman must be kept in dependence by
the males (of) their (families), and, if they attach
themselves to sensual enjoyments, they must be kept
under one's control.*

*Her father protects (her) in childhood, her husband
protects (her) in youth, and her sons protect (her) in old
age; a woman is never fit for independence."*

MANUSMRITI IX:17

*"(When creating them) Manu allotted to women (a love
of their) bed, (of their) seat and (of) ornament, impure
desires, wrath, dishonesty, malice, and bad conduct."*

MANUSMRITI IX:72

"Though (a man) may have accepted a damsel in due form, he may abandon (her if she be) blemished, diseased, or deflowered, and (if she have been) given with fraud."

VISHNUSMRITI XXV:14

"After the death of her husband, to preserve her chastity, or to ascend the pile after him."

MAHABHARATA,

ANUSASANA PARAVA, SECTION XXXVIII

"Rishi said, 'It is very true, O thou of slender waist! One incurs fault by speaking what is untrue. In saying, however, what is true, there can be no fault.' Thus addressed by him, the Apsara Panchachuda of sweet smiles consented to answer Narada's question. She then addressed herself to mention what the true and eternal faults of women are!' Panchachuda said, 'Even if high-born and endued with beauty and possessed of protectors, women wish to transgress the restraints assigned to them. This fault truly stains them, O Narada! There is nothing else that is more sinful than women. Verily, women, are the root of all faults. That is, certainly known to thee, O Narada! Women, even when

possessed of husbands having fame and wealth, of handsome features and completely obedient to them, are prepared to disregard them if they get the opportunity. This, O puissant one, is a sinful disposition with us women that, casting off modesty, we cultivate the companionship of men of sinful habits and intentions.

Women betray a liking for those men who court them, who approach their presence, and who respectfully serve them to even a slight extent. Through want of solicitation by persons of the other sex, or fear of relatives, women, who are naturally impatient of all restraints, do not transgress those that have been ordained for them, and remain by the side of their husbands. There is none whom they are incapable of admitting to their favours. They never take into consideration the age of the person they are prepared to favour. Ugly or handsome, if only the person happens to belong to the opposite sex, women are ready to enjoy his companionship. That women remain faithful to their lords is due not to their fear of sin, nor to compassion, nor to wealth, nor to the affection that springs up in their hearts for kinsmen and children. Women living in the bosom of respectable families envy the condition of those members of their sex that are young and well-adorned with jewels and gems and that lead a free life. Even those women that are loved by their husbands and treated with great respect, are seen to bestow their favours upon men that are hump-backed, that are blind, that are idiots, or that are dwarfs. Women may be seen to like the companionship of even those men that are destitute of the power of locomotion or those men that are endued with great ugliness of features. O great Rishi, there is no man in this world whom women may regard as unfit for companionship. Through inability to obtain persons of the opposite sex, or fear of relatives, or fear of death and imprisonment, women remain, of themselves, within the restraints prescribed for them. They are exceedingly restless, for they always hanker after new companions. In consequence of their nature being unintelligible, they are incapable of being kept in obedience by affectionate treatment. Their disposition is such that they are incapable of being restrained when bent upon transgression. Verily, women are like the words uttered by the wise. Fire is never satiated with fuel. Ocean can never be filled with the waters that rivers bring unto him. The Destroyer is never satiated with slaying even all living creatures. Similarly, women are never satiated with men. This, O celestial Rishi is another mystery connected with women. As soon as they see a man of handsome and charming features, unfailing signs of desire appear on their persons. They never show sufficient regard for even such husbands as accomplish all their wishes, as always do what is agreeable to them *and as protect them from want and danger. Women never regard so highly even articles of enjoyment in abundance or ornaments or other possessions of an agreeable kind as they do the companionship of persons of the opposite sex. The destroyer, the deity of wind, death, the nether legions, the equine mouth that roves through the ocean, vomiting ceaseless flames of fire, the sharpness of the razor, virulent poison, the snake, and Fire--all these exist in a state of union in women. That eternal Brahman whence the five great elements have sprung into existence, whence the Creator Brahma hath ordained the universe, and whence, indeed, men have sprung, verily from the same eternal source have women sprung into existence. At that time, again, O Narada, when women were created, these faults that I have enumerated were planted in them!'"*

ATHARVA VEDA 6:XI

"Asvattha on the Sami-tree. There a male birth is certified. There is the finding of a son: this bring we to the women-folk.

The father sows the genial seed, the woman tends and fosters it. This is the finding of a son: thus hath Prajapati declared.

Prajapati, Anumati, Sinivali have ordered it. Elsewhere may he effect the birth of maids, but here prepare a boy."

ATHARVA VEDA 14:I

"A wife is given by God to a husband to serve him and to bear him children. Further she is referred to by her husband as his subordinate and slave"

RIG VEDA 10:XCV:15

"Nay, do not die, Pururavas, nor vanish: let not the evil-omened wolves devour thee. With women there can be no lasting friendship: hearts of hyenas are the hearts of women."

Now let me bring up two events from the life of Jesus and Buddha.

There is a passage in the Gospel of John, that depicts some Pharisees bringing in a woman, who has committed adultery. As per the law made by Moses (which has nothing to do with God), the just thing to do is to stone the woman to death. Hence the following passage:

THE HEBREW BIBLE, DEUTERONOMY 22:22-24

"If a man be found lying with a woman married to an husband, then they shall both of them die, both the man that lay with the woman, and the woman: so shalt thou put away evil from Israel. If a damsel that is a virgin be betrothed unto an husband, and a man find her in the city, and lie with her; Then ye shall bring them both out unto the gate of that city, and ye shall stone them with stones that they die; the damsel, because she cried not, being in the city; and the man, because he hath humbled his neighbour's wife: so thou shalt put away evil from among you."

But, what's the point of reformation, if the old mistakes are repeated in the same pattern, instead of correcting them! When one says *"I am not come to destroy, but to fulfil"*, it simply means filtering the previous laws and getting rid of as much poison as possible (because there is no such thing as a perfect scripture). Hence the following passage of reformation:

JOHN 8:1-11

"Jesus went unto the mount of Olives. And early in the morning he came again into the temple, and all the people came unto him; and he sat down, and taught them. And the scribes and Pharisees brought unto him a woman taken in adultery; and when they had set her in the midst, They say unto him, Master, this woman was taken in adultery, in the very act. Now Moses in the law commanded us, that such should be stoned: but what sayest thou? This they said, tempting him, that they might have to accuse him. But Jesus stooped down, and with his finger wrote on the ground, as though he heard them not. So when they continued asking him, he lifted up himself, and said unto them, He that is without sin among you, let him first cast a stone at her. And again he stooped down, and wrote on the ground. And they which heard it, being convicted by their own conscience, went out one by one, beginning at the eldest, even unto the last: and Jesus was left alone, and the woman standing in the midst. When Jesus had lifted up himself, and saw none but the woman, he said unto her, Woman, where are those thine accusers? hath no man condemned thee? She said, No man, Lord. And Jesus said unto her, Neither do I condemn thee: go, and sin no more."

A similar event occurred in Buddha's life. When Buddha learned that King Pasenadi of Kosala was displeased that his queen had just given birth to a daughter rather than the desired son, he reassured the king as:

"A woman, O lord of the people, may turn out better than a man. She may be wise and virtuous, a devoted wife, revering her mother-in-law."

– SAMYUTTA NIKAYA 3:16

Here I must show you a glaring connection between Buddhism and Hinduism. The relation between Hinduism and what is called Buddhism at the present day is nearly the same as between Judaism and Christianity. Jesus Christ was a Jew, and Siddhartha Gautama (also known as Shakya Muni) was a Hindu. Upon attaining Nirvana, Siddhartha Gautama became the Buddha and started preaching a reformed version of the teachings from the ancient Indian scriptures. While on the other hand, around 500 years later, Jesus of Nazareth, under the influence of his transcendental experience, began teaching a reformed version of the Abrahamic faith.

All the religious faiths of human history were born out of human elements. Both good and bad elements of a person's personality take the form of sacred wish in the divine domain of transcendence. And if a person assimilates everything from the scriptures, without distinguishing the helpful elements from the harmful ones, it not only harms his or her own intellectual progress, but also brings chaos around that person. The ultimate consequence of such omnivorous attitude by the members of the most advanced species, is the doom of the whole mankind.

THREE

POLYGAMY

The path of evolution follows its own law, which is *natural selection*. It does not listen to you, me, or other biological humans, such as the religious giants. It is us the humans who must follow the path, or else we shall get extinct. And so did most of the prophets by giving in to nature, quite unconsciously. And being the slave of Mother Nature, most of the religious patriarchs adopted polygamy in the form of God's will, as per the law of wild evolution. Those male members of the species, simply fell prey to their own beastly instincts of spreading their genetic traits as widely as possible. It has nothing to do with God or spirituality. Rather it has all to do with survival of the species through procreation.

Then why the hell, are the religious laws of polygamy so much inclined towards men, and against women? Or let me put it this way. Why most of the religious scriptures glorify polygyny and not polyandry?

The answer to this question lies in the womb of Mother Nature. But before we go deep into Mother Nature's tummy to search for the answer, I must give you a little background on the term polygamy. The term *polygamy* is derived from the Greek word *polygamia,* which refers to the state of marriage to many spouses. There are mainly two types of polygamy – polygyny and polyandry. When a man is married to more than one wife at a time, it is called polygyny. And when a woman is married to more than one husband at a time, it is called polyandry.

Even though our modern human civilization has made this practice illegal throughout the world, many of the religious scriptures gloriously endorse the idea of polygamy. However, since the scriptures were developed from the standpoint of men's interest they don't really endorse the universal practice of polygamy, rather only one type of polygamy, that is polygyny. So, technically those scriptures permit a man to have more than one wife, but deem polyandry as a sin. Now, we shall explore the biological underpinning of such gender bias.

For a man, the optimal evolutionary strategy is to disseminate his genes as widely as possible, given his few minutes (or, alas, seconds) of investment in each encounter. While on the contrary, a woman invests a great deal of time and effort - a nine month long, risky, strenuous pregnancy, in each offspring. Naturally, over the course of millions of years, woman have evolved into a monogamous creature than man. Whereas the tendency of men is to be polygamous and promiscuous. Among the hundreds of human cultures throughout the world, only one, the Thodas of South India, have officially endorsed polyandry (the practice of having more than one husband or male mate).

Evolution 101: Men are more polygamous, women are more monogamous.

Now the question rises, if men are biologically more polygamous and women are more monogamous, then how can a romantic relationship ever last for long? The answer is again in the evolution of various brain regions. It is true that men will always be men with their innate wild attraction towards the large breasts and big hips around them even while being in a relationship. But through the process of Darwinian natural selection, an amazing brain region evolved, i.e. again the pre-frontal cortex, about which you learnt in the last chapter.

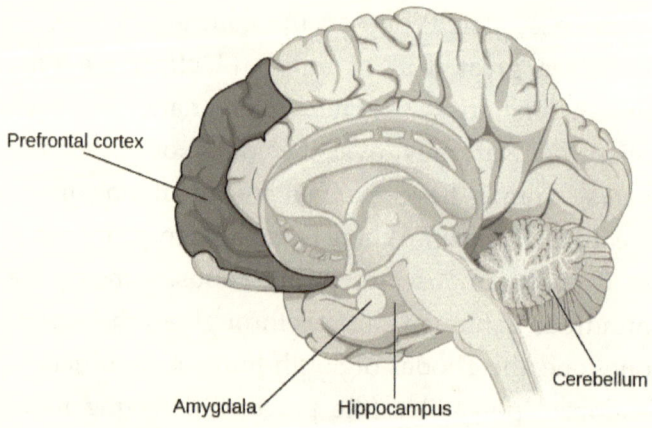

Figure 3.1 Prefrontal Cortex shown as darker region

Prefrontal cortex is the area of the brain that gives you the ability to keep all your momentary emotional impulses in check. So, even though a man is biologically incapable of stopping his testosterone level to go high when he visualizes a hot lady, he still can choose whether to act upon that momentary impulse of libido.

But in the altered state of transcendental consciousness the brain goes through a fascinating electrochemical storm, that defies all bounds and norms of the usual wakeful mind. Naturally, in such a state of mind, the beasts within the prophets were set lose, and got their wicked male urges inscribed into the scriptures, amidst the symposium of true philosophical teachings.

However, in the ordinary waking state of consciousness, the story is quite different. In fact, this is the story of a lucid mind, where the frontal lobes direct you towards the right path. Just imagine the concern that Mother Nature has for us. She put all her excellence in designing various brain circuits with utmost care so that we could keep our polygamous desires in check and lead a healthy, happy and abundant family life. She programmed us through selective pressure to go crazy with just a glance of our beloved ones.

In various studies we have found that merely having a look at the picture of your dear ones turns on the insula, anterior cingulate cortex, putamen, retrosplenial cortex and caudate, like a nuclear furnace. The insula and anterior cingulate cortex are typically associated with emotion oriented attentional states, whereas the retrosplenial cortex is involved in episodic memory recall, imagination, and planning for the future.

And off course the brain's love circuits share several brain regions with sexual arousal circuits. In several studies it has been found that certain regions do appear to be consistently heightened in response to sexual stimuli, such as the hypothalamus, putamen, visual cortex, inferior temporal cortex, orbitofrontal cortex, anterior cingulate cortex, parietal cortex, temporo-parietal junction, insula, ventral striatum, anterior temporal areas, amygdala, and basal ganglia. These studies have also pointed out a few regions that are distinctively active only in response to romantic stimuli. And those regions are caudate and ventral tegmental area. This significantly implies that the brain circuits of love and sexual arousal are anatomically distinctive yet intertwined.

But the craftsmanship of nature does not end just here. Like a nourishing mother, she embedded the ingredients of attachment right inside our head. Those ingredients are Oxytocin and Vasopressin. They play a critical role in forming a concept or perception of the partner whom we want to be with. They appear to build a strong profile of the mating partner through odor. The odor comes to be associated with a pleasurable and rewarding encounter with a particular partner. The same works in the visual domain. Oxytocin is not only responsible for the bonding of couples but also it is involved in maternal love towards a baby, whereas vasopressin is responsible for the commitment of the male towards his mate.

The influence of Oxytocin and Vasopressin is far more delicate than you can imagine. These two incredible hormones go to great lengths in order to keep us from being promiscuous. To illustrate this further, let me tell you the story of the prairie and the montane voles. It is a story of great biological interest. Among these two species, the prairie voles are mostly monogamous in nature, while the montane voles are promiscuous. Due to their brain circuits, the montane voles cannot maintain a healthy long-term relationship. If the release of Oxytocin and Vasopressin is blocked in prairie voles, they too become promiscuous. If however prairie voles are injected with these hormones but prevented from having sex, they will still continue to be faithful to their partners through a chaste monogamous relationship. That makes me wonder, what if we just inject the montane voles with Oxytocin and Vasopressin! Makes sense right!

ABHIJIT NASKAR

One might think that injecting the montane voles with these two hormones will somehow magically transform them into faithful monogamous creatures. But quite unfortunately it doesn't work that way. An injection of these love potions doesn't render them monogamous. Once secreted by the pituitary, these neurochemicals can only act if there are receptors for them in the brain. In the prairie voles there is an abundance of receptors for Oxytocin and Vasopressin in the reward centers of the brain. While on the contrary in the montane voles, receptors for these two hormones are not as abundant. Ergo, injecting the montane voles with excessive amounts of Oxytocin and Vasopressin doesn't make them monogamous, since there are not sufficient receptors for them in the reward centers of the brain.

There is a genetic cause behind this receptor variability. Prairie voles carry a longer version of the vasopressin receptor gene which makes them way more monogamous in behavior than the montane voles. Our two closest primate cousins, chimpanzees and bonobos also have different lengths of this gene, which match their social behaviors. Chimpanzees, who have the shorter gene, live in territorially based societies controlled

68

by males who make frequent, fatal war raids on neighboring troops. While on the other hand, Bonobos are run by female hierarchies and seal every social interaction with a bit of sexual impression. They are exceptionally social and have the long version of the gene. The human version of the gene is more like the bonobo gene. Differences in partner commitment may therefore be related to our individual differences in the length of this gene and in hormones.

So the ongoing joke among the women scientists is that the women should care more about the length of the vasopressin gene in their mates than about the length of anything else. Maybe someday there will be a drugstore test kit, similar to a pregnancy test, to know how long this gene is, so a woman can be sure she's getting the best guy before she commits. Just in a manner of speaking, male monogamy may therefore be somewhat predetermined for each individual and passed down genetically to the next generation. Possibly devoted fathers and faithful partners are born, not made or shaped by a father's example. However, as genetic engineering progresses, perhaps one day a woman will be able to alter the length of the vasopressin receptor gene in her man, to make a perfect faithful partner out of him.

The brain circuits for attachment have developed over the period of millions of years in order to serve the crucial evolutionary purpose of maintaining and promoting the survival of the species, while keeping the romantic partners faithful to each other. All the brain circuits were carefully crafted by Mother Nature with accurate precision so that bonding becomes a rewarding experience, without even the slightest need of having multiple mating partners.

One might wonder, why exactly such a common biological trait of pair-bonding evolved at all! Why not just stay in the primordial times and pleasure as many amorous encounters as possible! Why not squeeze all the juice out of life!

The answer lies in one of the major elements of life, i.e. the beautiful sensation of love or romance. Here comes the evolutionary reason behind the neuropsychological arrival of monogamy as a favored type of relationship. It is connected with the beautiful sensation of *love*. Evolutionarily speaking, love is all about procreation. And, only erotic passionate love making does not guarantee successful procreation. A lot more effort needs to be invested by the parents to actually ensure the survival of their progeny. For us humans, this can be up to twenty years. For this purpose, Mother Nature developed strategies beyond the *"one time fling"* approach to make mating partners collaborate until their progeny can survive on its own. Hence the neurological circuits of pair-bonding evolved. Along the way it led to the neuropsychological arrival of monogamy as a favored type of relationship, and left polygamy to the animals as a primordial practice.

Scientifically speaking, practicing and promoting polygamy in the name of religion in this evolved and civilized society is actually like signing the Declaration Certificate, that says:

"I hereby renounce my membership of mankind, since I am neither human nor kind. I declare that I no longer belong to the modern human species, i.e. the Homo sapiens. From now on I shall be counted among the swingers of the animal kingdom, such as the bonobo, spotted hyena or montane vole. I am simply an arrogant philandering barbarian."

However, I shall not go into the abyss of heinousness in a useless attempt of calling the religious leaders as barbarians, no matter the religion. It is not as simple as it seems. Being a biologist I can understand that it was very likely for them to fall for their mind's trickery, given their life conditions. But the conditions of humans have changed significantly in the last few centuries. Today, we have modern medicine, advanced technology, therapy and various other wonders of science. But back then, the only way to find solution to their life problems, was through introspection. And whenever, a human goes into the abyss of his or her own consciousness, often along with the precious pearls of wisdom, some molecules of poison bubble up to the surface.

Now I shall bring up some elements from the scriptures. You decide with your own intellectual abilities, whether they are the pearls of wisdom or the poison of savagery.

QURAN, SURAH 4 (AN-NISA), AYAT 1-3

"O mankind, fear your Lord, who created you from one soul and created from it its mate and dispersed from both of them many men and women. And fear Allah , through whom you ask one another, and the wombs. Indeed Allah is ever, over you, an Observer.

And give to the orphans their properties and do not substitute the defective [of your own] for the good [of theirs]. And do not consume their properties into your own. Indeed, that is ever a great sin.

And if you fear that you will not deal justly with the orphan girls, then marry those that please you of [other] women, two or three or four. But if you fear that you will not be just, then [marry only] one or those your right hand possesses. That is more suitable that you may not incline [to injustice]."

SAHIH AL-BUKHARI 7:62:2

"Narrated 'Ursa:

that he asked 'Aisha about the Statement of Allah: 'If you fear that you shall not be able to deal justly with the orphan girls, then marry (other) women of your choice, two or three or four; but if you fear that you shall not be able to deal justly (with them), then only one, or (the captives) that your right hands possess. That will be nearer to prevent you from doing injustice.' (4.3) 'Aisha said, "O my nephew! (This Verse has been revealed in connection with) an orphan girl under the guardianship of her guardian who is attracted by her wealth and beauty and intends to marry her with a Mahr less than what other women of her standard deserve. So they (such guardians) have been forbidden to marry them unless they do justice to them and give them their full Mahr, and they are ordered to marry other women instead of them.""

HEBREW BIBLE, DEUTERONOMY 21:15-17

"If a man have two wives, one beloved, and another hated, and they have born him children, both the beloved and the hated; and if the firstborn son be hers that was hated: Then it shall be, when he maketh his sons to inherit that which he hath, that he may not make the son of the beloved firstborn before the son of the hated, which is indeed the firstborn: But he shall acknowledge the son of the hated for the firstborn, by giving him a double portion of all that he hath: for he is the beginning of his strength; the right of the firstborn is his."

VISHNUSMRITI XXIV

"Now a Brâhmana may take four wives in the direct order of the (four) castes;

A Kshatriya, three;

A Vaisya, two;

A Sûdra, one only.

Among these (wives), if a man marries one of his own caste, their hands shall be joined.

In marriages with women of a different class, a Kshatriya bride must hold an arrow in her hand;

A Vaisya bride,. a whip;

A Sûdra bride, the skirt of a mantle.

No one should marry a woman belonging to the same Gotra, or descended from the same Rishi ancestors, or from the same Pravaras.

Nor (should he marry) one descended from his maternal ancestors within the fifth, or from his paternal ancestors within the seventh degree;

Nor one of a low family (such as an agriculturer's, or an attendant of the king's family);

Nor one diseased;

Nor one with a limb too much (as e. g. having six fingers),

Nor one with a limb too little;

Nor one whose hair is decidedly red;

Nor one talking idly.

There are eight forms of marriage

The Brâhma, Daiva, Ârsha, Prâgâpatya, Gândharva, Âsura, Râkshasa, and Paisâka forms.

The gift of a damsel to a fit bridegroom, who has been invited, is called a Brâhma marriage.

If she is given to a Ritvig (priest), while he is officiating at a sacrifice, it is called a Daiva marriage.

If (the giver of the bride) receives a pair of kine in return, a is called an Ârsha marriage.

(If she is given to a suitor) by his demand, it is called a Prâgâpatya marriage.

A union between two lovers, without the consent of mother and father, is called a Gândharva marriage.

If the damsel is sold (to the bridegroom), it is called an Âsura marriage.

If he seizes her forcibly, it is called a Râkshasa marriage. If he embraces her in her sleep, or while she is unconscious, it is called a Paisâka marriage.

Among those (eight forms of marriage), the four first forms are legitimate (for a Brâhmana);

And so is the Gândharva form for a Kshatriya.

A son procreated in a Brâhma marriage redeems (or sends into the heavenly abodes hereafter mentioned) twenty-one men (viz. ten ancestors, ten descendants, and him who gave the damsel in marriage).

A son procreated in a Daiva marriage, fourteen;

A son procreated in an Ârsha marriage, seven;

A son procreated in a Prâgâpatya marriage, four.

He who gives a damsel in marriage according to the Brâhma rite, brings her into the world of Brahman (after her death, and enters that world himself).

(He who gives her in marriage) according to the Daiva rite, (brings her) into Svarga (or heaven, and enters Svarga himself).

(He who gives her in marriage) according to the Ârsha rite, (brings her) into the world of Vishnu (and enters that world himself).

(He who gives her in marriage) according to the Prâgâpatya rite, (brings her) into the world of the gods (and enters that world himself).

(He who gives her in marriage) according to the Gândharva rite, will go to the world of Gandharvas.

A father, a paternal grandfather, a brother, a kinsman, a maternal grandfather, and the mother (are the persons) by whom a girl may be given in marriage.

On failure of the preceding one (it devolves upon) the next in order (to give her in marriage), in case he is able.

When she has allowed three monthly periods to pass (without being married), let her choose a husband for herself; three monthly periods having passed, she has in every case full power to dispose of herself (as she thinks best).

A damsel whose menses begin to appear (while she is living) at her father's house, before she has been betrothed to a man, has to be considered as a degraded woman: by taking her (without the consent of her kinsmen) a man commits no wrong."

No scripture is perfect. None of them come from any Supreme Creator, no matter what we call it. Jehovah, Brahman, Allah, God are simply a few among the countless terms the human brain constructs in an effort to explain the things it cannot understand so easily. The scriptures are just another proof that, how versatile the human brain is on this matter. It is not perfect, but it always thrives to become a better version of itself. Hence, occurred all the reformation time and again.

FOUR

HOMOSEXUALITY

I am ashamed to say that, in this so-called advanced civilization, there are still people who would be disgusted by the very utterance of the term *"homosexuality"*. And their only defense or justification for such disgust is: *"My scripture doesn't allow it, so it is a sin."*

Now one may ask for the sake of understanding, *what is a sin, what is not?*

To that person I say, *I am the least qualified person to answer that.* I am a Scientist, not a Magician or Theologian, who can just come up with any metaphysical answer that suits the purpose and feed it to the vulnerable masses. The entire concept of *sin,* is a sociological invention. It is one among the countless elements of social norms.

Any human action that goes against what is ordinary, is deemed as an anomaly, thus earns the title of a *sin*. From this perspective, homosexuality can indeed be hailed as an act of sin. But that is all primitive woo-woo stuff. It has nothing to do with the biological reality of life, as it is.

However, if you ask for an actual scientific answer, to whether Homosexuality is a sin, then, here it is. **Homosexuality is neither a sin, nor an anomaly. In biological terms, it is an evolutionary variation.**

For example, being a transgender is an evolutionary variation. It is neither an anomaly nor a disease. It may only be deemed by the ordinary human mind as an anomaly, because it is something out of the ordinary which the mind is usually accustomed to. The same goes for Homosexuals.

Life on earth is really wonderful. It is so vivid and versatile, that one who truly opens his or her eyes can't help being mesmerized by its vivacious beauty. Unaware of that beauty, human civilization has long perceived all biological variations as some of sort of alien infection. Homosexuality is the most stigmatized among them.

The psychological dynamics of heterosexual and homosexual relationships are just the same. Which means there is no difference in the true sensation of love between a heterosexual couple and a homosexual couple. The psychological adjustments made by the partners are just the same in both heterosexual and homosexual relationships. Even in terms of duration of the relationship, same-sex partners stay together for 20 years or even longer, just like heterosexual partners.

One may wonder what about parental capabilities!

Despite the baseless claims of those who oppose gay parents, no empirical study shows that having a gay male or lesbian parent is deleterious to children. Consequently, a growing number of courts have finally started to regard sexual orientation as irrelevant to a parent's ability to provide a good and healthy upbringing for his or her children.

In the 1990s, an unprecedented number of homosexual women and men chose to become parents in committed homosexual relationships. Many homosexual men and women had been parents before this time, but their children were usually conceived in a heterosexual marriage. Homosexual parents have often faced hostility from the conservative and apparently blind parts of the society, and have even been denied custody of their own children in many cases. This is what we call civilization!

Despite all our so-called advancements, only recently, people have started to think about the possibility that Homosexuality might not be an anomaly after all. In response to the intellectual vacuum created by the failures of psychiatry to solve the riddle of sexual orientation, we biologists recently embarked on the path of understanding the true biological basis of Homosexuality.

The term itself first appeared in German *"Homosexualität"* in a pamphlet published in Leipzig in 1869; and it entered the English language almost after two decades. We'll talk about the elements of homosexuality in the scriptures a little later. First, let's take a look at recent times. During the Middle Ages, engaging in sexual intercourse with a person of the same sex was regarded as a sin. Between the sixteenth and eighteenth century homosexual intercourse became a crime as well as a sin. But things got a little weird in the nineteenth century, when modern medicine and particularly the field of psychiatry (which was more about assumptions than actual science back then), came to view homosexuality as a form of mental illness. It was included in the first Diagnostic and Statistical Manual of the American Psychiatry Association, published in 1952. Labeling homosexuality as a form of psychopathology reflected nothing but psychiatrists' assumptions, derived from longstanding religious, cultural and legal traditions, and their clinical impressions of homosexuals who were seeking psychiatric treatment under the pressure of social norms.

For much of the twentieth century, even the scientific community deemed homosexuality as an aspect of psychopathic, paranoid, and schizoid personality disorders. Hence, the psychiatrists and other doctors of that time, made the *treatment* of homosexuality imperative. The medical profession loathed homosexuality to such an extent that virtually any proposed treatment seemed defensible. Lesbians were forced to submit to hysterectomies and estrogen injections, although it became clear that neither of these had any effect on their sexual orientation. Gay men were subjected to similar monstrosity. We lost many great minds in the hands of those ruthless and blind barbarians.

The most glorious name that comes to my mind is Alan Turing. He was a brilliant mathematician who conceived the modern digital computer. In 1950 he published an article entitled "Computing machinery and intelligence". In this paper he asks the question: Can machines think? Unable to define thought (like we are unable to define consciousness) he proposes what he calls an "Imitation game". It is played with three entities, a man (A), a computer (B), and an interrogator (C). The interrogator stays in a room apart. The objective of the game for the interrogator is to determine which of the other two is the man and which is the computer by asking questions. This imitation game is now called the Turing test and certain people believe it can help determine whether a computer is intelligent. No computer till this day has managed to fully fool the interrogator. Such was the genius of Alan Turing.

Yet he faced the most horrifying brutality of the so-called civilized world. He was a homosexual at the time when such an act was a criminal offence in the United Kingdom. Naturally, he was arrested and came to trial on 31 March 1952, after the police learned of his sexual relationship with a young Manchester man. He was convicted and given a choice between imprisonment and hormonal treatment to reduce his libido. Rather than going to prison, he accepted the option of treatment via injections of a synthetic oestrogen for the period of one year. This so-called treatment left its scars on Turing, rendering him impotent and caused gynaecomastia (a disorder of the endocrine system in which there is a non-cancerous increase in the size of breast tissue in males). While facing the relentless brutality of mankind, Turing already sensed his impending doom: *"no doubt I shall emerge from it all a different man, but quite who I've not found out."*

On June 8, 1954, Turing was found dead in his house by his housekeeper. He had died the day before of cyanide poisoning, a half-eaten apple beside his bed.

The lesson to learn from this dreadful event, is that in the face of human monstrosity, even the brilliant minds become helpless.

In the twentieth century, as psychiatrists made it mandatory to treat homosexuality, doctors tried every possible means, from castration to various kinds of aversion therapy, no matter how vicious. None of these could be shown to change the sexual orientation of a person whatsoever.

There was one specific scientist worth mentioning, since we are talking about sexual behavior. Among those who looked into the matter of sexual orientation a magnificent character was the sex researcher Alfred Kinsey, whose comprehensive reports, known as *Kinsey Reports (Sexual Behavior in the Human Male,1948 and Sexual Behavior in the Human Female,1953)* showed homosexuality to be surprisingly common across lines of family, class, and educational and geographic background. His research provoked a lot of controversy in the American society, as they essentially challenged the contemporary cultural values. Kinsey reports compelled the society to look at the reality of life, rather than the presumed illusory perfection.

Kinsey and his colleagues for many years attempted to find patients who had been converted from homosexual to heterosexual through therapy, and were surprised that they could not find even a single individual whose sexual orientation had been changed. When they interviewed individuals who claimed they had been homosexuals but were now functioning heterosexually, they found that all those persons were simply suppressing homosexual behavior and that they used homosexual fantasies to maintain potency when they attempted heterosexual intercourse.

Psychiatry proved to be nothing but a complete failure in showing that Homosexuality was a pathological condition that can be reversed. The first scientific study that broke the ongoing illusion of the psychiatrists that Homosexuality was a mental illness, was conducted by a young psychologist named Evelyn Hooker (yes, that was her name).

In 1956, Chicago, Hooker presented a study to a meeting of the American Psychological Association. She herself during her training routinely studied the so-called theory of homosexuality as a pathology. A group of young gay men with whom she had become friendly seemed, however, to be quite healthy and lucid in all daily activities. It suddenly appeared to Hooker that the scientific community still didn't know about Homosexuality. So, she received a study grant from the National Institute of Mental Health and chose a group of thirty gay men as the objects of her research and thirty straight men as controls; none of the sixty had ever sought or undergone psychiatric treatment. It was the first time homosexuals had been studied outside a medical setting or prison.

She conducted psychological tests on her sixty subjects, including the Rorschach ink-blot test, producing sixty psychological profiles. She removed all identifying marks, including those indicating sexual orientation. In order to eliminate her own biases, she gave them for interpretation to three eminent psychologists. One of them was Bruno Klopfer, who believed that he would be able to distinguish homosexuals from heterosexuals by means of the Rorschach test. However, quite astonishingly, none of the three could differentiate the homosexuals from the heterosexuals. In side-by-side comparisons of matched profiles, the heterosexuals and homosexuals were indistinguishable, demonstrating an equal distribution of pathology and mental health.

Hence, Hooker concluded from the study that homosexuality did not constitute a clinical entity and that it was not associated with pathology whatsoever. Her research was driven by her strong then-unconventional belief that for psychiatry to be minimally scientific, pathology must be defined in a way that is objective and empirically observable.

Her findings were subsequently replicated in numerous empirical studies of both women and men. The weight of growing empirical evidence, coupled with changing social norms and the development of politically active gay community in the United States of America, compelled the Directors of the American Psychiatric Association to officially remove homosexuality from the Diagnostic and Statistical Manual, in 1973.

The movement to declassify homosexuality as a diagnosis has been strongly supported by the American Psychological Association (APA.) ever since 1974. APA has passed numerous legal resolutions to support equal right for lesbians and gay men in employment, child custody and access to services.

However, there is still a huge difference between theory and practice. Humans shall always remain humans, no matter the position of science. Ever since 1973, the scientific approach among the mainstream psychotherapists has been to help the homosexual clients adjust successfully to their sexual orientation and live life to the fullest. Despite all this, some (non-)psychotherapists and religious counselors continue to make disgraceful attempts to convert homosexuals into heterosexuals.

The point is, five decades of psychiatric evidence clearly demonstrates that homosexuality is immutable, irreversible and nonpathological. On top of that, we recently started to gather a growing body of empirical data implicating the biological foundation in the development of sexual orientation.

After conducting relentless neurobiological studies, on the biological foundation of sexual orientation, we have been able to move Homosexuality from the domain of psychiatric illnesses into the realm of normal variants of human sexual behavior.

Sexual orientation of a person gets biologically imprinted in the brain circuits during early foetal development. Prenatal sex hormones directly influence the development of the neural network of sexual orientation. Prenatal exposure to an opposite-sex hormonal environment, leads the brain circuits to develop homosexual orientation. This prenatal hormonal environment also has ever-lasting effects on the individual's all kinds of behavioral traits. And once imprinted into the neural map, sexual orientation, as well as all other behavioral traits of a person are absolutely irreversible and indelible.

Now, let me elucidate on how the prenatal hormonal environment leaves an irreversible imprint on a person's sexual orientation.

During the early foetal development, if a genetically female brain is exposed to higher levels of testosterone, a male sex hormone, it leads the entire nervous system including the brain circuits to develop along more male-typical lines. While on the other hand, if a genetically male brain is exposed to higher estrogen, a female sex hormone, it leads the entire nervous system including the brain circuits to develop along more female-typical lines. Also, several family and twin studies provide clear evidence for a genetic component to both male and female sexual orientation.

The complex and fascinating wiring of the human brain for sexual orientation occurs during fetal development, following the blueprint of that individual's genes and sex hormones. The behavioral expression of his or her brain wiring will then be influenced and shaped by environment and culture. Thus, environmental and social norms can influence a person's identity, but they cannot alter something so rudimentary – the biological blueprint of sexual orientation.

The biological manifestation of homosexual blueprint can be observed in different brain structures of homosexual men and women. Take the anterior commissure for example. It is the bundle of high-speed fibers, that connects the two hemispheres of the brain. The functional aspect of this structure is that it is involved in various brain processes related to cognitive abilities and language. In straight women, it is anatomically larger than in straight men. Such sexual dimorphism gives straight women better cognitive abilities and verbal fluency than straight men.

But, things gets interesting when it comes to homosexual brain. The sexual dimorphism of anterior commissure seen in heterosexual brains, is reversed in homosexual brains. Hence, between a homosexual man and woman, it is the man that has better cognitive abilities and verbal fluency than the homosexual woman. Also we can say that a gay man has better verbal fluency than a straight man, and a straight woman has better verbal fluency than a lesbian.

Another fascinating anatomical difference between the homosexual brain and heterosexual brain, lies in the connectivity of amygdala, the fear response center of the brain. Brain scan studies have shown that the connectivity of the amygdala in the gay male brain is more like that of the straight female brain than of the straight male brain. Whereas, in the lesbian brain it is more like that of the straight male brain than of the straight female brain.

The sexually dimorphic wirings of amygdala leads to better startle response in straight women, than straight men. But, in case of homosexual wirings of the amygdala, the story is exactly the opposite. Lesbians have a lower startle response than straight women, but in a similar range to straight men. Whereas, gay men have more startle response than straight men, in a similar range to straight women.

All these studies are eloquent proof that homosexuality has nothing to do with pathology, but it has all to do with biology. Biologically speaking, prenatal opposite-sex hormonal exposure and genetic variation lead to homosexual traits. It is simply a biological variation of human sexual behavior. So, definitely it is not an anomaly or a disease.

We the neuroscientists have spent decades trying to understand the process of sexual orientation, and the biological functions underneath it. It is really a fundamental concept that must not be taken lightly. All our countless experiments, observations and examinations of the human biology, lead to one simple conclusion – *Homosexuality is not an abnormality.* Now you can look back at the scriptures which were written hundreds and thousands of years ago by ordinary humans and ask yourself does it make sense at all!

THE HEBREW BIBLE, LEVITICUS 18:22

"You shall not lie with a male as one lies with a female; it is an abomination."

THE HEBREW BIBLE, LEVITICUS 20:13

"If a man also lie with mankind, as he lieth with a woman, both of them have committed an abomination: they shall surely be put to death; their blood shall be upon them."

QURAN, SURAH 7 (AL-A'RAF), AYAT 80-84

"And [We had sent] Lot when he said to his people, "Do you commit such immorality as no one has preceded you with from among the worlds? Indeed, you approach men with desire, instead of women. Rather, you are a transgressing people." But the answer of his people was only that they said, "Evict them from your city! Indeed, they are men who keep themselves pure." So We saved him and his family, except for his wife; she was of those who remained [with the evildoers]. And We rained upon them a rain [of stones]. Then see how was the end of the criminals."

SUNAN ABU DAWUD 38:4447

"Narrated by Abdullah ibn Abbas: The Prophet (peace be upon him) said: If you find anyone doing as Lot's people did, kill the one who does it, and the one to whom it is done."

SAHIH AL-BUKHARI 7:72:774

"Narrated by Abdullah ibn Abbas: The Prophet cursed effeminate men; those men who are in the similitude (assume the manners of women) and those women who assume the manners of men, and he said, "Turn them out of your houses." The Prophet turned out such-and-

such man, and 'Umar turned out such-and-such woman."

MANUSMRITI VIII:369,370

"A damsel who pollutes (another) damsel must be fined two hundred (panas), pay the double of her (nuptial) fee, and receive ten (lashes with a) rod.

But a woman who pollutes a damsel shall instantly have (her head) shaved or two fingers cut off, and be made to ride (through the town) on a donkey."

Scriptures are simply books with philosophical ideas, mostly perceived in a transcendental state of consciousness. Those ideas are all product of mankind's own consciousness. Naturally, like all other creations of the human mind, some of those ideas can smoothen the path of intellectual development and lead humanity towards a better future, while some others can completely take away our true humanity and throw it into the Grand Canyon of chaos.

FIVE

TERRORISM

All the discussions, observations, investigations and examinations of religion, come down to one simple question: *does religion cause violence?*

To the untrained eye, the answer would probably appear as a *yes*. And this is exactly why all the anti-religious scholars of our time have landed on the bestseller list, in a brainless pursuit of blaming the religions for destructive events, despite the fact that they cannot cite actual empirical evidence which even mildly suggests that religion is hazardous to human health.

If you ask whether the scriptures contain sacred verses that actually command a person to act in a destructive manner, then the answer would again be *yes*. But those verses were the product of the prophets' brains' fight-or-flight response. And, there comes a huge difference between theory and practice. A person with a lucid mind would never go to such lengths to actually act on those verses literally. Only individuals with an unstable mental health take those violent religious creeds as their last resort of absolution.

Figure 5.1 No matter how you pray, it still fills your brain with stress-relieving chemicals (Source: Autobiography of God: Biopsy of A Cognitive Reality)

Countless experiments of modern neuroscience, eloquently depict, in usual circumstances of daily life, religions, or more specifically religious rituals have great health impact over the human biology. Only a fool with no scientific knowledge would deny it.

If so, then how is it possible, that all the destructive events of human history took place in the name of religion! *Holy War*, they call it.

(Here I am using the term *"Holy War"*, as a universal term to refer to any kind of violent actions in the name of religion, no matter the religion)

Alongside being the smartest species on earth, we are also one of the most aggressive ones that have ever existed. We are the Tyrannosaurus Rex of the mammals. As a species we have killed every single animal form that we have encountered. Every single discovery we have made from gunpowder to atomic fission has been to kill a perceived enemy. Since the year 1500, approximately 150 million human beings like you and me, have died during armed conflicts between nations.

Throughout history various cultures around the world have raped, slaughtered and drowned the planet in blood, in the names of Jehovah, Allah, Rama or the Great Cosmic Guide. And like always, the so-called intellectuals, only look at the surface of the problem and make their easy conclusions. Sociologists and historians with their limited thinking attribute these episodes to political or economic causes. Whereas, anti-religious scholars, simply blame it on the religions. But all these conclusions are simply based on assumptions, opinions and hatred.

It is easy to make conclusions. It is the brain's innate instinct to make conclusions that suit the personal needs, based on available environmental information. And in this case the only available information are the terrorizing actions of some individuals and their apparently direct connection with a few verses from the scriptures. And, as far as your brain is concerned, it just has to come up with a plausible enough explanation of things that you don't understand, with the use of conjectures, beliefs and gut feelings. Your brain doesn't give a damn about whether the conclusion it makes for you is actually really. The only concern of your brain is to satisfy your curiosity. The conclusion must seem plausible enough to you, and only you. It is all personal. And when the available environmental information are almost the same throughout the whole world, it is not a surprise, the majority of the human brains on earth come to the same conclusion – *religion causes terrorism.*

Now comes the reality, which is not as simple as it seems to the majority of the humans. The problem lies much deeper into the human mind, inside the limbic system to be specific.

The human biology has a neurological predisposition to act in profoundly hostile ways. And at the root of *terrorism*, there is this neurological predisposition of hostility at play. It is the biological mechanism of fight-or-flight response, that has evolved through millions of years. And the only difference between us, the so-called lucid humans and those who fight the *Holy War,* is that we have a functional Prefrontal Cortex, that keeps our wild and hostile instincts in check. But unfortunately, the warriors of the Holy War have very little activity in their PFC, that is caused by their boiling rage against the society.

However, fuel for terrorism comes from the creeds of the religious organizations that fundamentally depict that there is only one absolute and undeniable truth, and all others even mildly different truths are expendable. And when mentally unstable individuals join hands and get hold of such creeds, they unconsciously foster a dividing line between them and the rest of the world. Their "us versus them" mentality neurologically overloads their brain's limbic system. This generates fear and hostility toward people who hold different beliefs.

The moment they feel like their personal belief is threatened and they need to defend it, their brain kicks into overdrive, generating stress chemicals that put them into a state of fight-or-flight.

The fight-or-flight response begins in the amygdala, which triggers a neurological response in the hypothalamus. This initial response is followed by activation of the pituitary gland and secretion of the Adrenocorticotropic hormone (ACTH) by the anterior pituitary gland. The adrenal gland is activated almost simultaneously and releases the neurochemical epinephrine, which you may know as adrenalin. The release of these chemicals triggers the production of the stress hormone cortisol in the adrenal gland. Cortisol increases blood pressure, blood sugar, and suppresses the immune system.

The initial response and subsequent reactions are triggered in an effort to create a boost of energy. This boost of energy is activated by epinephrine binding to liver cells and the subsequent production of glucose. Additionally, the circulation of cortisol functions to turn fatty acids into available energy, which prepares the muscles throughout the body for physical response. Catecholamines, such as epinephrine (adrenalin) and norepinephrine (noradrenalin), facilitate immediate physical reactions associated with a preparation for violent outbursts.

In such a mental state, individuals fostering the hostile mentality towards the world, become engulfed with anger, which is the most primitive of all emotions. It is also the most difficult one to tame. And with the arrival of anger, the human mind is overwhelmed with the sensation of defensiveness, anxiety and aggression. Anger ultimately shuts down the functioning of the frontal lobes, which is actually the completely opposite of what happens when you practice religious and spiritual rituals. As a result those mentally unstable individuals who are already boiling with rage against all other belief systems in the world, lose all touch with natural healthy reality. They don't just lose the ability to be rational, but also the awareness that they are being irrational.

With their frontal lobes significantly inactive, they lose all biological capability to listen to another person. They lose their ability to empathize with people coming from a different belief system. They feel like the whole world has gone against them. The end product is their catastrophic violent outburst over the whole world. Through their violent eruptions, their boiling rage triggers a cascade of neurochemicals that destroys their brain-regions of emotional responsiveness permanently.

It is like, the so-called terrorists are at the mercy of their own brain. The entire mental process behind their acts of violence takes place like this.

First, circumstances pour the minds of those individuals with hatred and rage towards the society.

And when that pain, hatred, and rage become unbearable, they turn to the scriptures as the final resort, in a pursuit to find absolution.

They take comfort in the violent elements of the sacred texts, because that's what they are searching for. Apparently the sacred commands from God to fight for religion and slaughter whoever holds a different belief system than them, make complete sense.

QURAN, SURAH 2 (AL-BAQARAH), AYAT 190-194

"Fight in the way of Allah those who fight you but do not transgress. Indeed. Allah does not like transgressors. And kill them wherever you overtake them and expel them from wherever they have expelled you, and fitnah is worse than killing. And do not fight them at al-Masjid al- Haram until they fight you there. But if they fight you, then kill them. Such is the recompense of the disbelievers. And if they cease, then indeed, Allah is Forgiving and Merciful. And if they cease, then indeed, Allah is Forgiving and Merciful. Fight them until there is no [more] fitnah and [until] worship is [acknowledged to be] for Allah . But if they cease, then there is to be no aggression except against the oppressors. [Fighting in] the sacred month is for [aggression committed in] the sacred month, and for [all] violations is legal retribution. So whoever has assaulted you, then assault him in the

same way that he has assaulted you. And fear Allah and know that Allah is with those who fear Him."

QURAN, SURAH 2 (AL-BAQARAH), AYAT 216

"Fighting has been enjoined upon you while it is hateful to you. But perhaps you hate a thing and it is good for you; and perhaps you love a thing and it is bad for you. And Allah Knows, while you know not."

QURAN, SURAH 3 (ALI-'IMRAN), AYAT 55-56

"[Mention] when Allah said, "O Jesus, indeed I will take you and raise you to Myself and purify you from those who disbelieve and make those who follow you [in submission to Allah alone] superior to those who disbelieve until the Day of Resurrection. Then to Me is your return, and I will judge between you concerning that in which you used to differ. And as for those who disbelieved, I will punish them with a severe punishment in this world and the Hereafter, and they will have no helpers.""

QURAN, SURAH 3 (ALI-'IMRAN), AYAT 141

"And that Allah may purify the believers [through trials] and destroy the disbelievers."

QURAN, SURAH 3 (ALI-'IMRAN), AYAT 150-152

"But Allah is your protector, and He is the best of helpers. We will cast terror into the hearts of those who disbelieve for what they have associated with Allah of which He had not sent down [any] authority. And their refuge will be the Fire, and wretched is the residence of the wrongdoers. And Allah had certainly fulfilled His promise to you when you were killing the enemy by His permission until [the time] when you lost courage and fell to disputing about the order [given by the Prophet] and disobeyed after He had shown you that which you love. Among you are some who desire this world, and among you are some who desire the Hereafter. Then he turned you back from them [defeated] that He might test you. And He has already forgiven you, and Allah is the possessor of bounty for the believers."

QURAN, SURAH 4 (AN-NISA), AYAT 74-76

"So let those fight in the cause of Allah who sell the life of this world for the Hereafter. And he who fights in the cause of Allah and is killed or achieves victory - We will

bestow upon him a great reward. And what is [the matter] with you that you fight not in the cause of Allah and [for] the oppressed among men, women, and children who say, "Our Lord, take us out of this city of oppressive people and appoint for us from Yourself a protector and appoint for us from Yourself a helper?" Those who believe fight in the cause of Allah , and those who disbelieve fight in the cause of Taghut. So fight against the allies of Satan. Indeed, the plot of Satan has ever been weak."

QURAN, SURAH 4 (AN-NISA), AYAT 87-96

"Allah - there is no deity except Him. He will surely assemble you for [account on] the Day of Resurrection, about which there is no doubt. And who is more truthful than Allah in statement. What is [the matter] with you [that you are] two groups concerning the hypocrites, while Allah has made them fall back [into error and disbelief] for what they earned. Do you wish to guide those whom Allah has sent astray? And he whom Allah sends astray - never will you find for him a way [of guidance]. They wish you would disbelieve as they disbelieved so you would be alike. So do not take from among them allies until they emigrate for the cause of Allah . But if they turn away, then seize them and kill

them wherever you find them and take not from among them any ally or helper. Except for those who take refuge with a people between yourselves and whom is a treaty or those who come to you, their hearts strained at [the prospect of] fighting you or fighting their own people. And if Allah had willed, He could have given them power over you, and they would have fought you. So if they remove themselves from you and do not fight you and offer you peace, then Allah has not made for you a cause [for fighting] against them. You will find others who wish to obtain security from you and [to] obtain security from their people. Every time they are returned to [the influence of] disbelief, they fall back into it. So if they do not withdraw from you or offer you peace or restrain their hands, then seize them and kill them wherever you overtake them. And those - We have made for you against them a clear authorization. And never is it for a believer to kill a believer except by mistake. And whoever kills a believer by mistake - then the freeing of a believing slave and a compensation payment presented to the deceased's family [is required] unless they give [up their right as] charity. But if the deceased was from a people at war with you and he was a believer - then [only] the freeing of a believing slave; and if he was from a people with whom you have a treaty - then a compensation payment presented to his family and the freeing of a believing slave. And whoever does not find [one or cannot afford to buy one] - then [instead], a fast for two months consecutively, [seeking] acceptance of repentance from Allah . And Allah is ever Knowing and Wise. But whoever kills a believer intentionally - his recompense is Hell, wherein he will abide eternally, and Allah has become angry with him and has cursed him and has prepared for him a great punishment. O you who have believed, when you go forth [to fight] in the cause of Allah , investigate; and do not say to one who gives you [a greeting of] peace "You are not a believer," aspiring for the goods of worldly life; for with Allah are many acquisitions. You [yourselves] were like that before; then Allah conferred His favor upon you, so investigate. Indeed Allah is ever, with what you do, Acquainted. Not equal are those believers remaining [at home] - other than the disabled - and the mujahideen, [who strive and fight] in the cause of Allah with their wealth and their lives. Allah has preferred the mujahideen through their wealth and their lives over those who remain [behind], by degrees. And to both Allah has promised the best [reward]. But Allah has preferred the mujahideen over those who remain [behind] with a great reward - Degrees [of high position] from Him and forgiveness and mercy. And Allah is ever Forgiving and Merciful."

Quran, Surah 4 (An-Nisa), Ayat 104

"And do not weaken in pursuit of the enemy. If you should be suffering - so are they suffering as you are suffering, but you expect from Allah that which they expect not. And Allah is ever Knowing and Wise."

Quran, Surah 5 (Al-Ma'idah), Ayat 33-34

"Indeed, the penalty for those who wage war against Allah and His Messenger and strive upon earth [to cause] corruption is none but that they be killed or crucified or that their hands and feet be cut off from opposite sides or that they be exiled from the land. That is for them a disgrace in this world; and for them in the Hereafter is a great punishment, Except for those who return [repenting] before you apprehend them. And know that Allah is Forgiving and Merciful."

Quran, Surah 8 (Al-Anfal), Ayat 12

"[Remember] when your Lord inspired to the angels, "I am with you, so strengthen those who have believed. I will cast terror into the hearts of those who disbelieved, so strike [them] upon the necks and strike from them every fingertip.""

SAHIH AL-BUKHARI 1:35

"The person who participates in (Holy Battles) in Allah's cause and nothing compels him do so except belief in Allah and His Apostle, will be recompensed by Allah either with a reward, or booty (if he survives) or will be admitted to Paradise (if he is killed)."

SAHIH AL-BUKHARI 8:387

"Allah's Apostle said, "I have been ordered to fight the people till they say: 'None has the right to be worshipped but Allah'. And if they say so, pray like our prayers, face our Qibla and slaughter as we slaughter, then their blood and property will be sacred to us and we will not interfere with them except legally.""

SAHIH AL-BUKHARI 52:177

"Allah's Apostle said, "The Hour will not be established until you fight with the Jews, and the stone behind which a Jew will be hiding will say. "O Muslim! There is a Jew hiding behind me, so kill him.""

THE HEBREW BIBLE, DEUTERONOMY 13

"If there arise among you a prophet, or a dreamer of dreams, and giveth thee a sign or a wonder, And the sign or the wonder come to pass, whereof he spake unto thee, saying, Let us go after other gods, which thou hast not known, and let us serve them; Thou shalt not hearken unto the words of that prophet, or that dreamer of dreams: for the LORD your God proveth you, to know whether ye love the LORD your God with all your heart and with all your soul. Ye shall walk after the LORD your God, and fear him, and keep his commandments, and obey his voice, and ye shall serve him, and cleave unto him. And that prophet, or that dreamer of dreams, shall be put to death; because he hath spoken to turn you away from the LORD your God, which brought you out of the land of Egypt, and redeemed you out of the house of bondage, to thrust thee out of the way which the LORD thy God commanded thee to walk in. So shalt thou put the evil away from the midst of thee. If thy brother, the son of thy mother, or thy son, or thy daughter, or the wife of thy bosom, or thy friend, which is as thine own soul, entice thee secretly, saying, Let us go and serve other gods, which thou hast not known, thou, nor thy fathers; Namely, of the gods of the people which are round about you, nigh unto thee, or far off from thee, from the one end of the earth even unto the other end of the earth; Thou shalt not consent unto him, nor hearken unto him; neither shall thine eye pity him,

neither shalt thou spare, neither shalt thou conceal him: But thou shalt surely kill him; thine hand shall be first upon him to put him to death, and afterwards the hand of all the people. And thou shalt stone him with stones, that he die; because he hath sought to thrust thee away from the LORD thy God, which brought thee out of the land of Egypt, from the house of bondage. And all Israel shall hear, and fear, and shall do no more any such wickedness as this is among you. If thou shalt hear say in one of thy cities, which the LORD thy God hath given thee to dwell there, saying, Certain men, the children of Belial, are gone out from among you, and have withdrawn the inhabitants of their city, saying, Let us go and serve other gods, which ye have not known; Then shalt thou inquire, and make search, and ask diligently; and, behold, if it be truth, and the thing certain, that such abomination is wrought among you; Thou shalt surely smite the inhabitants of that city with the edge of the sword, destroying it utterly, and all that is therein, and the cattle thereof, with the edge of the sword. And thou shalt gather all the spoil of it into the midst of the street thereof, and shalt burn with fire the city, and all the spoil thereof every whit, for the LORD thy God: and it shall be an heap for ever; it shall not be built again. And there shall cleave nought of the cursed thing to thine hand: that the LORD may turn from the fierceness of his anger, and shew thee mercy, and have compassion upon thee, and multiply thee, as he hath sworn unto thy fathers; When thou shalt hearken to the voice of the LORD thy God, to keep all his commandments which I command thee this day, to do that which is right in the eyes of the LORD thy God."

Through the sacred verses filled with violence and self-righteousness, the minds of the angry individuals find a way to get rid of all their misery. At that unstable state of consciousness, they are drawn to the description of the *Holy War*. They visualize a glimmer of hope. They feel absolutely immersed in it. Finally when they emerge as *holy warriors,* they are no longer humans, from the emotional perspective. They emerge as wild beasts, neurologically almost unable to feel human emotions, like empathy, love, kindness and compassion. Consequently the whole world faces the wrath of the most primitive of all human elements in the name of God's judgment.

If we look closely, we shall discover the ghost from our own past in those holy warriors. They are nothing but the remnants of our times gone by. In the primitive days of humanity, all of us were exactly like them. Let me elucidate on this matter.

Humanity is a hodge-podge of various characteristic features, some of which are primitive, and some are relatively modern. Evolution has endowed us with these fascinating behavioral traits. Our limbic brain houses all our emotions, both positive and negative. It also houses the monster that had to fight for its survival every single moment of its existence. Over time we learnt to tame that monster. For this purpose we slowly developed the precious layers of cerebral cortex, that eventually shoved the monster deep down at the abyss of human consciousness. The layers of the cortex also made us the smartest species on earth. Whereas, the monster remains down there, sleeping at the depth of our unconscious mind. But the protective layers of the cortex that we have developed over millions of years act as a binding spell and keep the monster from lashing out to the surface of consciousness.

Hence, I'd say, terrorism is simply an evolutionary vestigial trait of our species, - a horrible reminder to our primeval past.

SIX

FUNDAMENTALISM

The greatest threat to mankind is not terrorism, rather it is *fundamentalism.* Fundamentalism acts as a mainstream fuel for terrorism. The moment, fundamentalism disappears from the face of earth, society would finally get rid of religious terrorism. Terrorism feeds on fundamentalism. But I won't go to such barbaric lengths in saying, the fundamentalists deserve to be killed, because if I do, then there would be no difference between those ancient apes and us the scientists.

The term "fundamentalism" was first coined in 1920 by C. L. Laws in the Baptist Watchman-Examiner in order to describe a Christian movement formed to oppose liberal theology in America. The movement had become known through a series of books under the title *The Fundamentals*. Eventually, fundamentalism turned out to be widely characterized as strict literalism of any scripture, be it New Testament, or the Quran or any other.

The definition itself is out of place in this 21st century, and absolutely against world harmony. If one starts accepting and obeying all the words from the scriptures literally, it would only cause chaos. For instance, if someone reads the sacred words of *Jihad* from the Quran and starts obeying them literally, what would be the result of such obedience? He would start killing all the *kafirs*, i.e. all the human beings who are not Muslims. Now if someone says to him *"this is wrong"*, chances are, he would not waste a single moment to decapitate that person, because according to Mohammed's transcendental state of consciousness, by opposing a *Muslim* one is actually opposing Allah.

Here the problem is not religion, rather it is fundamentalism, i.e. literal interpretation of scriptures. Such literalism only creates separation among humans. Anything that promotes such separation, creates a cognitive impediment in the development of the species.

In this context, I'd like to point out something elementary. Fundamentalism and Terrorism may be intertwined with each other, but they are completely different things.

While fundamentalism hails both good and bad elements of the scriptures, as divine will, Terrorism on the hand, takes place when circumstances force mentally unstable individuals cling to the negative aspects of the scriptures.

Fundamentalism acts as a major fuel in fostering terrorism. Without fundamentalism, terrorism won't exist in such an extreme form. It might have made sense back in the days, when there was no such thing as science. But now we know more than enough about our own existence, to at least stop letting the extremely harmful elements of scriptures contaminate human life and society.

It is true, that compared to what there is to be known in the universe, what we do know is minuscule. Yet, in the last few centuries, we have truly stepped into the domain of extraordinary scientific advancement. In a planet which is around 4.5 billion years old, we the humans, i.e. Homo sapiens evolved from our ancestor species Homo heidelbergensis only 200,000 years ago. And in the entire period of our existence, it is only in the last several centuries, that we have truly transformed from a developing species, into the most intelligent one on this planet.

Plato, Socrates, Aristotle and many more great minds laid the groundwork for the development of modern science. Over the foundation of philosophy, history witnessed the daring ventures of human excellence by both philosophical and scientific geniuses, such as Leonardo-da-Vinci, Copernicus, Galileo, Kepler, Bacon, Darwin, Newton and so on. And the chain of reaction they triggered with their extraordinarily abnormal thinking, given their surrounding ignorance and fundamentalism, resulted into the evolution of our modern science.

Today, every scientific method seeks to explain all events of Mother Nature, but in a reproducible way. We scientists don't have the luxury to make claims, based only on assumptions. Even when we propose a thought experiment or a hypothesis on a certain phenomenon, we reach to that conclusion only through strenuous examination of all the facts available concerning that phenomenon. Then that falsifiable (a statement which has an inherent possibility, that it can be proven false) hypothesis is further tested through countless experiments by others scientists. It is only through all these rigorous observations and experiments, that a hypothesis is proved. However, if a hypothesis turns out to unsatisfactory, it is then either modified or disproved. Modification or disproof of a hypothesis, is evidence of our progress. Only if the hypothesis survives the rigorous testing, we adopt it into the framework of a scientific theory.

For example, *Biological Evolution* was first proposed by late eighteenth and nineteenth century naturalists, as a *hypothesis* or a *theory*, based on their observation of the striking similarities between Neanderthal fossils and Homo sapiens skeleton. Through further rigorous testing of countless fossil records *evolutionary biology* or *Darwinism* became an established natural phenomenon. Today, when we use the phrase *"theory of evolution"*, we actually refer to the established fact of biological evolution of life on planet earth. Here the term *"theory"* is just a habitual recurrence.

The first fully formed theory of evolution, was proposed by a French Naturalist Jean-Baptiste Lamarck (1744-1829) in the early nineteenth century, as *the theory of the transmutation of species*. Then in 1858 Charles Darwin and Alfred Russel

Wallace published their conception of evolution based on the process of natural selection.

Later, in his book, "On the Origin of Species by Means of Natural Selection" published in 1859, Charles Darwin proposed what is now called the *theory of evolution*. This book, along with its best-known companion "The Descent of Man and Selection in Relation to Sex", published in 1871, called for a major change in scientific thinking about the origin of life, particularly in the field of biology. Darwin was not the first scientist to propose a theory of evolution nor was he the foremost thinker on the subject in 1859. He was simply the product of his time.

When Darwin conceived the idea of "natural selection" soon after his return from the voyage of the Beagle in 1838, he set himself to the task of streamlining the arguments and collecting supporting evidence. Although his ideas were well known among English biologists, Darwin hesitated to publish them in fear of the reaction of the religious fundamentalist society and the Church, and he continued to accumulate an increasingly impressive body of data consistent with his theory. Darwin's hand was forced when another biologist Alfred Russel Wallace independently conceived the same idea, and mailed it in a concise ten-page letter to him. Darwin quickly wrote up a short outline, and both papers were presented to the Linnaean Society simultaneously in 1858. Had Wallace bypassed Darwin and published first, right now we would be speaking of Wallace's theory of evolution by natural selection, and Darwin would remain a poorly known documenter of Wallace's theory. Darwin, of course, is a household name. It is a pity that Wallace is almost completely unknown except among biologists and historians of science, since he was an equally brilliant scholar and independently came up with the same idea.

However, instead of disputing endlessly over priority, as many of today's scientists do, Darwin and Wallace cheerfully acknowledged each other's contributions and Wallace even wrote a book called Darwinism, advocating what he referred to as "Darwin's" theory of natural selection. Upon hearing about this book, Darwin responded: *"You should not speak of Darwinism for it can as well be called Wallacism."*

Darwin was a very clever scientist. His use of the research of fellow scientists is one of the most fascinating features of his book "The Origin of Species". Eventually, this book became the bedrock of biology, and the term *Darwinism* became unavoidably intertwined with our very existence.

Upon the arrival of Darwinism, the Christian fundamentalists, based on their literal interpretation of the Genesis creation narrative, constructed an illusory conflict between science and religion. This conflict created by the fundamentalists concerning evolution, was actually about their personal irrational obedience to their book. It had nothing to do with religion.

However, this worthless battle between evolution and creationism, is relatively recent compared to the one I am now going to tell you about. This conflict goes way back at the time of the origin of modern astronomy. Today the whole world knows that the earth and all other planets in our solar system revolve around the sun. This astronomical model is called Heliocentrism. All these have been learnt by the human species through constant experimentation and observation. But this scientific model of modern astronomy didn't get established overnight. When first proposed, Heliocentrism was about to break a predominant spell of the early model of astronomy, i.e. Geocentrism in which the sun and all other planets were presumed to be revolving around the earth. Naturally, the path ahead was tremendously rusty.

Now, let's turn the clock back and witness how exactly modern astronomy evolved from a primitive society.

The night sky has always been a subject of human curiosity. The dark ocean above our head filled with bright little spots has fascinated many early human civilizations on earth. Babylonians, Egyptians, Greeks, and Indians all had a fascination for the celestial objects and the elite of the intellectuals of those civilizations built theories to explain the miracles of the heavens. And the most advanced among them were the Greeks and the ancient Indians.

Earlier all those celestial objects were accepted to be from the gods. Nobody felt the push to truly understand them, because it was not necessary. All the questions concerning *why* and *how* led to one answer *"God has made it this way"*.

Eventually the explanation of the universe took more logical and scientific forms. However, it was not until the Greeks' development, that proper theories about the earth and the revolution of the planets emerged. The most predominant theory of the structure of the universe in the ancient world was the geocentric model, called Geocentrism, which says that the earth is at the center of the universe, and every other celestial body revolves around the earth.

The origin of this theory is obvious; it is the elementary naked eye observation of the movement of the objects in the sky. The path of an object in the sky always seems to be in the same vicinity and repeatedly it rises from the east and sets in the west approximately at the same points on the horizon. Also, the earth always seems to be stationary or motionless and still. Therefore, the closest conclusion anyone can make is that these objects move in circles around our own earth. Greeks were strong supporters of this theory.

In the 4th century BC, two of the greatest Greek philosophers, Plato and his student Aristotle, wrote works based on the geocentric model. However, it was not until 2nd Century AD, that Geocentrism was standardized.

In the 2nd Century Greco-Egyptian astronomer Claudius Ptolemaeus assimilated centuries of works on geocentrism and developed the first fully formed geocentric model of the universe. He presented the first ever standardized model of geocentrism in his main astronomical work Almagest. And even after Ptolemy's death, the Ptolemaic system lasted for more than 2000 years unchallenged, as it was perfectly compatible with the contemporary religious conservative society.

At such circumstances, in the sixteenth century, Nicolaus Copernicus came up with a new idea. He hypothesized that all of the planets traveled around the sun. This is called a heliocentric model (Helios is Greek for Sun). However, you should note that it wasn't an entirely new concept. Some ancient Greeks and Indians, such as the Greek philosopher Aristarchus of Samos (310-230 BC) and the Indian mathematician and astronomer Aryabhatta (476-550 AD), had speculated that the Sun was at the center of our solar system.

Still, it was a controversial idea, because of its inherent scientific elements that essentially contradicted the Bible and the accepted traditions of that time. During that Renaissance era, Nicolaus Copernicus published his mathematical model of the universe *"De Revolutionibus Orbium Coelestium"* (*On The Revolution of the Celestial Spheres, here the term *"revolution"* refers to Copernicus' challenge to the contemporary religious authoritarianism).

Copernicus began to write it in 1506 and finished it in 1530, but did not publish it until the year of his death, in 1543. Although he was in good standing with the Church and had dedicated the book to Pope Paul III, the published form contained an unsigned preface by Andreas Osiander, a German Lutheran Theologian defending the system and arguing that it was useful for computation even if its hypotheses were not necessarily true. Because of that preface, the Copernicus' book invoked very little debate on whether it might be heretical during the next 60 years.

However, the air of suggestion among Dominicans (the Roman Catholic Order of Preachers), that the teaching of heliocentrism should be banned, was growing dense. In fact, some years after the publication of Copernicus' book John Calvin (a principal figure in the development of the system of Christian theology later called Calvinism) preached a sermon in which he denounced those who *"pervert the course of nature"* by saying that *"the sun does not move and that it is the earth that revolves and that it turns"*.

Eventually all this hatred from the Catholic fundamentalist society was poured upon a glorious character of our scientific history called Galileo.

This is the person who had to face the most brutal consequences of being different in a conservative society.

Galileo Gelilei was born in 1564 in Pisa, Italy. He is often called *the father observational astronomy*. His job as a professor of astronomy at University of Padua in Italy required Galileo to teach the theory of that time, which said that the universe was geocentric. It was the most comfortable idea back then.

In 1609 he heard about the invention of a telescope in Holland. It was being used to watch for ships coming into ports there. From a simple description, he made a far better telescope for himself. He had heard about Copernicus' heliocentric theory many years before he ever looked through the telescope. And when Galileo pointed his telescope to the skies, the observations he made, convinced him of the truth of heliocentric model.

When he turned his new-fangled telescope at the night sky, the first object he observed was Venus, the brightest planet in the sky. To the unaided eye, Venus just looks like a bright point of light. But through his telescope, Galileo could see the disk of Venus and found that it did not look same as the weeks and months went by. Sometimes, Venus showed just a thin crescent, while at other times much more of it was lit up. It showed phases, just like the Moon. He was the first person to see the phases of Venus and the moons of planet Jupiter. Studying all observations, he concluded that the heliocentric model is the true model of our solar system.

In 1610, Galileo published his book *Sidereus Nuncius (Starry Messenger)*, a book that one cannot even dream of in the fundamentalist environment of that time. The Starry Messenger described all the surprising observations that Galileo had made with the new telescope, namely the phases of Venus and the Galilean moons of Jupiter. With these observations he promoted the heliocentric theory of Nicolaus Copernicus (published in De Revolutionibus Orbium Coelestium in 1543).

In February 1615, prominent Dominicans brought Galileo's work on heliocentrism to the attention of the Inquisition, because they appeared to violate the Holy Scripture. And naturally due to the sheer rigidity of heinous fundamentalism, his discoveries were met with brainless opposition.

In February 1616, the Inquisition assembled a committee of theologians, known as qualifiers, who delivered their unanimous report condemning heliocentrism as *"foolish and absurd in philosophy, and formally heretical since it explicitly contradicts in many places the sense of Holy Scripture"*.

All the books containing heliocentric ideas were banned including Copernicus' De Revolutionibus Orbium Coelestium and Galileo was ordered *"to abstain completely from teaching or defending this doctrine and opinion or from discussing it... to abandon completely... the opinion that the sun stands still at the center of the world and the earth moves, and henceforth not to hold, teach, or defend it in any way whatever, either orally or in writing."*

Late in 1632 Galileo published a book named *"Dialogo sopra i due massimi sistemi del mondo"* (Dialogue Concerning the Two Chief World Systems), as an account of conversations among a Copernican scientist, Salviati, an impartial and witty scholar named Sagredo, and a ponderous Aristotelian named Simplicio, who employed stock arguments in support of egocentricity. This book brought him a lot of fame, but also he lost many of his defenders in Rome.

Naturally, in the year 1633 he was ordered to stand trial once again on suspicion of heresy, *"for holding as true the false doctrine taught by some that the sun is the center of the world"*. It was pretty obvious, since he was previously commanded by the Inquisition in 1616, to refrain from all works concerning the doctrine of heliocentrism.

Ultimately sheer stupidity hailed Galileo as guilty, and the sentence of the Inquisition was issued on 22 June 1633. The sentence consisted of three essential parts:

1. Galileo was found *"vehemently suspect of heresy,"* namely of having held the opinions that the Sun lies motionless at the center of the universe, that the Earth is not at its center and moves, and that one may hold and defend an opinion as probable after it has been declared contrary to Holy Scripture. He was required to *"abjure, curse, and detest"* those opinions.

2. He was sentenced to formal imprisonment at the pleasure of the Inquisition. On the following day this was commuted to house arrest, which he remained under for the rest of his life.

3. His offending Dialogue was banned; and in an action not announced at the trial, publication of any

of his works was forbidden, including any he might write in the future.

Even though Galileo was locked away by the idiotic fundamentalist society of that time, it couldn't lock away the knowledge he shared. In the hands of other daring scientists, such as Sir Isaac Newton, Johannes Kepler, and many others, heliocentrism eventually found its deserving place into the human conscience of the society. As a result in the mid-eighteenth century the religious opposition of the theory of heliocentrism started to fade.

The imprisonment of Galileo is simply a reminder to, how far a person can go in the pursuit of science.

Back in those days, when science was at its cradle, our perception of the universe was mostly dependent on our interpretation of the scriptures, which we thought to be a golden mine of actual scientific information given by God. There was no reason to not believe any single word from them. In fact, we were not even looking for actual answers, but we were looking for signs of divine intervention in everything around us. We relied primarily upon our beliefs. And this belief constructed all those conflicts between the newly developing scientific minds and the religious orthodox society. To be honest, there was no conflict between science and religion ever. The conflicts were actually between two different systems of human understanding – one was science, which was based on rigorous observations and examinations, and the other was fundamentalism, that's based on undisputed belief on the scriptures.

The conflicts were simply between the personal conservative ideas of the fundamentalists and the dawn of human conscience. And it was pretty obvious that those two systems of understanding were not going to embrace each other so easily. It takes time. And now is the time, that I can truly say, **there can never be a conflict between science and religion, once you understand the spiritual knack of the human brain circuits.**

Today, Science of the Human Mind - Neuroscience explains the true nature of Religion and Spirituality. We neuroscientists have found that inside the human brain, science and religion co-exist quite harmoniously. They are unique yet compatible with each other. And this harmony has evolved through millions of years of constant struggle for survival. Today the human brain is capable of abstract thinking as well as complex associations and analyses. It is the organ of unparalleled importance. It is the organ of creation. It is the organ with which you see, perceive, observe, think, recollect, reminiscence, pleasure, love, kiss, make love and even make babies. It is the organ that created the scriptures. And ultimately, it is the organ that creates your perception of God, as an anxiolytic system.

The Brain creates what the Human needs. Religion is an evolutionary need of an ordinary human being. To fulfill that need, the brain constructed the Holy Books of history. And anything that a human creates has flaws in it. No human creation is the embodiment of perfection. And if you don't judge a human creation with your own conscience, just because it is ancient, then the flaws in it would lead you away from the path of humanity.

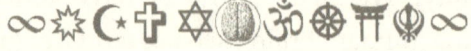

SEVEN

THE FATE OF RELIGIONS

The purpose of religion, by its core, is freedom of the human mind. Any religion that tends to bind a person with the twisted knots of textual fanaticism, needs reformation. You can read as many books of religion as you want. But in the beginning of mankind, religion was born as a natural sensation from a beautiful neural interplay within the brain. So, to know religion you must feel it in your heart. Books only give you a bunch of vocabulary. But all the words in the world are created by humans, so are the books and so are the religions. And anything that humans create has both personal and universal interest in it. It's up to you, which side of interest you want to choose.

First you must be clear on the idea of religion. If to you religion means reading books and obeying every single word from it without a slightest bit of reasoning, then such perception would only bring destruction upon you and the world. Also there are people who use the words from those books to justify their own filthy actions. Let's take a conservative Muslim, for example. Say, the conservative Muslim male Homo sapiens (I won't call such creature a human, regardless of the religion, since his action here shows no sign of humanity) is found to be beating his wife. Now, if someone says to him *"this is wrong"*, he would naturally reply, *"this is a divine thing to do, my book says so"*. Now, if a Christian says *"my book is older, so you should stop obeying your book and start obeying mine"*, there will come the Buddhist, and say, *"my book is much older still, obey mine"*. Then will come the Jew, and say, *"my book is even older, so just follow mine"*. And in the end will come the Hindu and say *"my books are the oldest of all, obey them"*. Therefore referring to books will only make a mess of the mankind and tear the species into pieces.

Biologically, the true essence of religion, lies within the human mind, not in any book, no matter how old they are. Religion is not a book. All the current dilemma of the human society whether a specific *religion* (supposedly Islam) *is a religion of peace or of violence,* is founded upon the belief that a religion is a specific book. But the reality is not so simple. Religion is a natural feature of the human mind that created all the scriptures in the world, not the other way around. Religion is the mind's urge to become better in the spiritual domain. All those books only depict that urge. They only show what a handful of individuals in the history of mankind experienced when their brain made them transcend from an ordinary wakeful state into an ultimate state of divinity and bliss. Thus all divinity, spirituality and religiosity exist within your very brain.

It appears, the whole universe exists in your own human consciousness, which happens to be the most advanced among all species on planet earth. Bring that 3 pound spongy grey matter into use. Human brain holds all the answers to you questions. You just need to ask the right question at the right moment. It is not about what the books permit you to do, or not to do. It is about what you as a precious member of mankind permit yourself to do. It is about finding the kingdom of conscience.

Any idea of separation is a bondage. True liberation of the mind is in non-differentiation. We cannot change the words of the scriptures, some of which came from minds in transcendence, some from selfishness. It is no use in changing the scriptures in the pursuit of reforming religion. It is the human perception of religion, that needs reformation, not religion itself. For that, you must stop accepting those words of the dead people without truly seeing them. Do not accept a single word from the scriptures without properly examining it, no matter the religion. Be fearless and ignore the ordinary critics as worms. Be brave and upright. Embrace the goodness around you. Taste the love of Christmas, the radiance of Diwali, the brotherhood of Ramadan, the feast of Sukkot and assimilate anything that appeals to you.

Whatever may be the position of science, whatever may be the position of philosophy, as long as there is such a thing as death in the world, as long as there is such a thing as weakness in the human heart, as long as the human heart sheds a drop of tear in weakness, there shall be a faith in God and divinity. So, use this opportunity to spread the most common, yet most vivacious and universal elements of all religions – *Love* and *Compassion.*

Let's break the barriers down and embrace each other. Let's build a world beyond chaos and religious conflicts, where the human mind does not allow the poison of fundamentalism invade the society. Let's construct a world better than this one, where one's individuality finds new meaning in accepting the goodness of others. And the effort it takes to build that world, is minuscule. You just need to open your eyes and perceive the new planet. Perceive the prolific land of universal tolerance, where there is no authoritarianism of any single scripture or religious institution, and it shall become a reality.

I dream of a planet where the science of the mind, brings the Bible, the Vedas, the Quran, and all other scriptures together and binds them with the golden twine of harmony. In that world of one humanity all humans are my brothers, sisters and friends.

Bibliography

Abram Hoffer and Humphrey Osmond, The Hallucinogens (New York: Academic Press, 1967).

Abetti, Giorgio (2012). "Cosmology". Encyclopedia Americana (Online ed.). Grolier.

Allen, L.S., Gorshi, R.A., 1992. Sexual orientation and the size of the anterior commissure in the human brain, Proc. Natl Acad. Sci. USA 89

Ashbrook, James, and Carol Albright. The Humanizing Brain: Where Religion and Neuroscience Meet. Cleveland, OH: Pilgrim Press, 1997.

Andresen, Jensine, and Robert Forman, eds. Cognitive Models and Spiritual Maps. Bowling Green, Ohio: Imprint Academic, 2000.

Azari, Nina, Janpeter Nickel, Gilbert Wunderlich, Michael Niedeggen, Harald Hefter, Lutz Tellmann, Hans Herzog, Petra Stoerig, Dieter Birnbacher, and Rudiger Seitz. "Neural Correlates of Religious

Experience." European Journal of Neuroscience 13, no. 8 (2001)

Anton, S. C. Natural history of Homo erectus. American Journal of Physical Anthropology S37, (2003)

Alemseged, Z., Spoor, F., Kimbel, W.H., Bobe, R., Geraads, D., Reed, D., Wynn, J.G., 2006. A juvenile early hominin skeleton from Dikika, Ethiopia. Nature 443.

Asfaw, B., White, T., Lovejoy, O., Latimer, B., Simpson, S., Suwa, G., 1999. Australopithecus garhi: a new species of early hominid from Ethiopia. Science 284.

Armstrong E (1982) Mosaic evolution in the primate brain: differences and similarities in the hominoid thalamus. In: Armstrong E, Falk D (eds) Primate brain evolution: methods and concepts. Plenum, New York

Ardila A, Gomez J. Paroxysmal "feeling of somebody being nearby". Epilepsia 1988

Babayev ES, Allahverdiyeva AA. Effects of geomagnetic activity variations on the physiological and psychological state of functionally healthy humans: some results of Azerbaijani studies. Adv Space Res 2007.

Baars, B. (1988), A Cognitive Theory of Consciousness (New York: Cambridge University Press).

Bailey, J.M., 2003. The Man Who Would be Queen. Joseph Henry Press, Washington, DC.

Bailey, J.M., Pillard, R.C., 1995. Genetics of human sexual orientation. Ann. Rev. Sex. Res. 60.

Bailey, J.M., Barbow, P., Wolfe, M., Mikach, S., 1995. Sexual orientation of adult sons of gay fathers. Dev. Psychol. 31.

Bailey, J.M., Pillard, R.C., Dawood, K., Miller, M.B., Farrer, L.A., Trivedi, S., Murphy, R.L., 1999. A family history study of male sexual orientation using three independent samples. Behav. Genet. 29.

Bailey, J.M., Dunne, M.P., Martin, N.G., 2000. Genetic and environmental influences on sexual orientation and its correlates in an Australian twin sample. J. Pers. Soc. Psychol. 78.

Blanchard, R., 2004. Quantitative and theoretical analyses of the relation between older brothers and homosexuality in men. J. Theor. Biol. 230.

Blanchard, R., Bogaert, A.F., 1996. Homosexuality in men and number of older brothers. Am. J. Psychiatry 153.

Blanchard, R., Ellis, L., 2001. Birth weight, sexual orientation and the sex of preceding siblings. J. Biosoc. Sci. 33.

Bancaud, J., Brunet-Bourgin, F., Chauvel, P. & Halgren, E. (1994), 'Anatomical origin of deja vu and vivid "memories" in human temporal lobe epilepsy', Brain, 117.

Bear, D.M. (1979), 'Personality changes associated with neurologic lesions', in Textbook of Outpatient Psychiatry, ed. A. Lazare (Baltimore, MD: Williams and Wilkins Co.).

Bogen,J.E.(1995a), 'On the neurophysiology of consciousness: Part I. An overview', Consciousness and Cognition, 4.

Bogen, J.E. (1995b), 'On the neurophysiology of consciousness: Part II. Constraining the semantic problem', Consciousness and Cognition, 4.

Buxhoeveden DP, Switala AE, Litaker M, Roy E, Casanova MF (2001) Lateralization of minicolumns in human planum temporale is absent in nonhuman primate cortex. Brain Behav Evol 57

Bregman, A. (1981), 'Asking the "what for" question', in Perceptual Organization, ed. M. Kubovy & J. Pomerantz (Hillsdale, NJ: Lawrence Erlbaum Associates).

Blumenschine, R. J. et al. Late Pliocene Homo and hominid land use from Western Olduvai Gorge, Tanzania. Science 299, (2003)

Berger, T., Trinkaus, E., 1995. Patterns of trauma among the Neandertals. Journal of Archaeological Science 22

Bobe, R., Behrensmeyer, A.K., 2004. The expansion of grassland systems in Africa in relation to mammalian evolution and the origin of the genus Homo. Palaeogeography, Palaeoclimatology, Palaeoecology 207

Bickerton, D. (2009). Adam's tongue: How humans made language and how language made humans. New York: Hill and Wang.

Bjorklund, D. F., and K. Kipp (1996). "Parental investment theory and gender differences in the evolution of inhibition mechanisms." Psychol Bull 120 (2)

Blair, R. J., J. S. Morris, et al. (1999). "Dissociable neural responses to facial expressions of sadness and anger." Brain 122 (Pt. 5).

Brothers, L. (2002). The social brain: A project for integrating primate behavior and neurophysiology in a new domain. In J. T. Cacioppo et al. (Eds.), Foundations in neuroscience. Cambridge, MA: MIT Press.

Bradley, M. M., M. Codispoti, et al. (2001). "Emotion and motivation II: Sex differences in picture processing." Emotion 1 (3).

Bradley, M. M., B. Moulder, et al. (2005). "When good things go bad: The re- flex physiology of defense." Psychol Sci 16 (6).

Brebner, J. (2003). "Gender and emotions." Personality and Individual Differences 34.

Bremner, J. D., R. Soufer, et al. (2001). "Gender differences in cognitive and neural correlates of remembrance of emotional words." Psychopharmacol Bull 35 (3).

Briton, N. J., and J. A. Hall (1995). "Beliefs about female and male nonverbal communication." Sex Roles 32.

Bhagavad-gita As It Is, The Bhaktivedanta Book Trust 2010.

Blanke, O. and Arzy, S. "The Out-of-Body Experience: Disturbed Self-Processing at the Temporo-Parietal Junction" THE NEUROSCIENTIST 2005.

Blanke O, Ortigue S, Landis T, Seeck M. 2002. Stimulating illusory own-body perceptions. Nature.

Bonda E, Petrides M, Frey S, Evans A. 1995. Neural correlates of mental transformations of the body-in-space. Proc Natl Acad Sci U S A.

Breedlove, SM. 1992. Sexually dimorphism in the vertebrate nervous system, The Journal of Neuroscience 12.

Brugger P, Agosti R, Regard M, Wieser HG, Landis T. 1994. Heautoscopy, epilepsy, and suicide. J Neurol Neurosurg Psychiatr.

Brugger P, Regard M, Landis T. 1997. Illusory reduplication of one's own body: phenomenology and classification of autoscopic phenomena. Cogn Neuropsychiatr.

Belisheva, N. K., Popov, A. N., Petukhova, N. V., Pavlova, L. P., Osipov, K. S., Tkachenko, S.E., & Baranova, T.I. (1995). Quantitative and qualitative evaluations of the effect of geomagnetic variations on the functional state of the brain. Biophysics, 40

Booth, J. N., Koren, S. A., & Persinger, M. A. (2005). Increased feelings of the sensed presence and increased geomagnetic activity at the time of the experience during transcerebral exposure to weak complex magnetic fields. International Journal of Neuroscience, 115

Beauregard, Mario, and Denyse O'Leary. The Spiritual Brain. New York: HarperCollins, 2007.

Beauregard, Mario, and Vincent Paquette. "Neural Correlates of a Mystical Experience in Carmelite Nuns." Neuroscience Letters 405, no. 3 (2006)

Beauregard, Mario, Jerome Courtemanche, and Vincent Paquette. "Brain Activity in Near-Death Experiencers during a Meditative State." Resuscitation 80, no. 9 (2009)

Benson, Herbert. Timeless Healing: The Power and Biology of Belief. New York: Scribner, 1996

Blackmore, Susan. Dying to Live: Science and Near-Death Experiences. London: HarperCollins, 1993

Boyer, Pascal. Religion Explained. New York: Basic Books, 2002

Buss, D. (1990). "International preferences in selecting mates: A study of 37 cultures." Journal of Cross-Cultural Psychology 21.

Buss, D. D. (2003). Evolutionary Psychology: The New Science of Mind, 2nd ed. New York: Allyn & Bacon. Buss, D. M. (1989). "Conflict between the sexes: Strategic interference and the evocation of anger and upset." J Pers Soc Psychol 56 (5).

Buss, D. M. (1995). "Psychological sex differences. Origins through sexual selection." Am Psychol 50 (3).

Buss, D. M. (2002). "Review: Human Mate Guarding." Neuro Endocrinol Lett 23 (Suppl 4).

Buss, D. M., and D. P. Schmitt (1993). "Sexual strategies theory: An evolutionary perspective on human mating." Psychol Rev 100 (2).

Byne, W., Lasco, M.S., Kemether, E., Shinwari, A., Edgar, M.A., Morgello, S., Jones, L.B., Tobet, S., 2001. The interstitial nuclei of the human anterior hypothalamus: an investigation of variation with sex, sexual orientation and HIV status. Horm. Behav. 40.

Camperio-Ciani, A., Corna, F., Capiluppi, C., 2004. Evidence for maternally inherited factors favouring male homosexuality and promoting female fecundity. Proc. Royal Soc. London B Biol. Sci. 271.

Cantor, J.M., Blanchard, R., Paterson, A.D., Bogaert, A.F., 2002. How many gay men owe their sexual orientation to fraternal birth order?. Arch. Sex. Behav. 31.

Chivers, M.L., Rieger, G., Latty, E., Bailey, J.M., 2004. A sex difference in the specificity of sexual arousal. Psychol. Sci. 15.

Chung, W.C.J., Swaab, D.F., De Vries, G.J., 2000. Apoptosis during sexual differentiation of the bed nucleus of the stria terminalis in the rat brain. J. Neurobiol. 43.

Clark, M.M., Robertson, R.K., Galef Jr.., B.G., 1996. Effects of perinatal testosterone on handedness of gerbils: support for part of the Geschwind–Galaburda hypothesis. Behav. Neurosci. 110.

Cherry, N. (2002). Schumann resonances, a plausible biophysical mechanism for the human health effects of solar/geomagnetic activity. Natural Hazards,26

Cook, C. M., & Persinger, M. A. (1997). Experimental induction of the "sensed presence" in normal subjects and an exceptional subject. Perceptual and Motor Skills, 85

Cooke, Roger (1997). The History of Mathematics: A Brief Course. Wiley-Interscience.

Cook, C. M., & Persinger, M. A. (2001). Geophysical variables and behavior: XCII. Experimental elicitation of the experience of a sentient being by right hemispheric, weak magnetic fields: interaction with temporal lobe sensitivity. Perceptual and Motor Skills, 92

Classical Hindu Mythology: A Reader in the Sanskrit Puranas, Temple University Press 1978.

Comparative primate energetics and hominid evolution. American Journal of Physical Anthropology 102.

Churchland, P.S. (1986), Neurophilosophy (Cambridge, MA: The MIT Press).

Churchland, P.S. & Ramachandran, V.S. (1993), 'Filling in: Why Dennett is wrong', in Dennett and His Critics: Demystifying Mind, ed. B. Dahlbom (Oxford: Blackwell Scientific Press).

Churchland, P.S., Ramachandran, V.S. & Sejnowski, T.J. (1994), 'A critique of pure vision', in Large- scale Neuronal Theories of the Brain, ed. C. Koch & J.L. Davis (Cambridge, MA: The MIT Press).

Cicurel, R., "L'ordinateur ne digérera pas le cerveau", Sarina Editions, 2013

Crick, F. (1994), The Astonishing Hypothesis: The Scientific Search for the Soul (New York: Simon and Schuster). Crick, F. (1996), 'Visual perception: rivalry and consciousness', Nature, 379.

Crick, F. & Koch, C. (1992), 'The problem of consciousness', Scientific American, 267.

Crowe, Michael J. (1990). Theories of the World from Antiquity to the Copernican Revolution. Mineola, NY: Dover Publications.

Cardena, Etzel, Steven Lynn, and Stanley Krippner, eds. Varieties of Anomalous Experience: Examining the Scientific Evidence. New York: American Psychological Association, 2004

d'Aquili, Eugene. "Human Ceremonial Ritual and the Modulation of Aggression." Zygon 20, no. 1 (1985) d'Aquili, Eugene. "Senses of Reality in Science and Religion." Zygon 17, no 4 (1982)

d'Aquili, Eugene. "The Biopsychological Determinants of Religious Ritual Behavior." Zygon 10, no. 1 (1975)

d'Aquili, Eugene. "The Myth-Ritual Complex: A Biogenetic Structural Analysis." Zygon 18, no. 3 (1983)

d'Aquili, Eugene, and Andrew Newberg. The Mystical Mind: Probing the Biology of Religious Experience. Minneapolis: Fortress Press, 1999.

d'Aquili, Eugene, Charles Lauglin, and John McManus. The Spectrum of Ritual: A Biological Structural Analysis. New York: Columbia University Press, 1979

Dawood, K., Pillard, R.C., Horvath, C., Revelle, W., Bailey, J.M., 2000. Familial aspects of male homosexuality. Arch. Sex. Behav. 29.

Drake, Stillman (1978). Galileo At Work. Chicago: University of Chicago Press.

Dreyer, J.L.E. (1953), A History of Astronomy from Thales to Kepler, New York, NY: Dover Publications.

Dickson, N., Paul, C., Herbison, P., 2003. Same-sex attraction in a birth cohort: prevalence and persistence in early adulthood. Soc. Sci. Med. 56.

DuPree, M.G., Mustanski, B.S., Bocklandt, S., Nievergelt, C., Hamer, D.H., 2004. A candidate gene study of CYP19 (aromatase) and male sexual orientation. Behav. Genet. 34.

Deacon, Terrance. The Symbolic Species: The Co-Evolution of Language and the Brain. New York: Norton, 1998

Dennett, Daniel. Breaking the Spell: Religion as a Natural Phenomenon. New York: Penguin, 2007.

Dewhurst, Kenneth, and A. W. Beard. "Sudden Religious Conversions in Temporal Lobe Epilepsy." British Journal of Psychiatry 117 (1970)

Duke University's Center for Spirituality, Theology, and Health. "Latest Religion and Health Research Outside Duke." http://www.spiritualityandhealth.duke.edu/resources/pdfs/Research%20-%20latest%20 outside%20Duke.pdf.

Darwin, Charles. "On the origin of species by means of natural selection" (original edition, 1859).

Darwin, Charles. "The Descent of Man" (original edition, 1871).

Dawkins, Richard. "The God Delusion", Bantam Press, 2006

Daly DD. 1958. Ictal affect. Am J Psychiatry.

Decety J, Sommerville JA. 2003. Shared representations between self and other: a social cognitive neuroscience view. Trends Cogn Sci.

Denning TR, Berrios GE. 1994. Autoscopic phenomena. Br J Psychiatry.

Devinsky O, Feldmann E, Burrowes K, Bromfield E. 1989. Autoscopic phenomena with seizures. Arch Neurol.

Downing PE, Jiang Y, Shuman M, Kanwisher N. 2001. A cortical area selective for visual processing of the human body. Science.

Dewhurst K, Beard AW. Sudden religious conversions in temporal lobe epilepsy. 1970 Epilepsy Behav 2003

Dastur H.M., Desai A.D., "A comparative study of brain tuberculomas and gliomas based upon 107 case records of each". Brain. 1965

Devinsky O, Lai G. Spirituality and religion in epilepsy. Epilepsy Behav 2008.

Dostoyevsky, The Possessed (original edition 1872)

Dostoyevsky, The Brothers Karamazov, (original edition 1880)

Dostoyevsky, The Insulted and Injured, (original edition 1861)

Dostoyevsky, The Idiot, (original edition 1869)

Dennett, D.C. (1991), Consciousness Explained (Boston, MA: Little, Brown and Co.).

Devinsky, O., Morrell, MJ, Vogt, BA. (1995) 'Contribution of anterior cingulate cortex to behavior', Brain, 118.

Domínguez-Rodrigo, M., Pickering, T.R., Semaw, S., Rogers, M.J., 2005. Cutmarked bones from Pliocene archaeological sites at Gona, Afar, Ethiopia: Implications for the functions of the world's oldest stone tools. Journal of Human Evolution 48, 109-121

DeGiorgio, M. et al. Out of Africa: modern human origins special feature: explaining worldwide patterns of human genetic variation using a coalescent-based serial founder model of migration outward from Africa. PNAS USA 106, 16057-16062 (2009)

Delson, E., Harvati, K., 2006. Return of the last Neanderthal. Nature 443.

Edelman, G. M. (1992). Bright air, brilliant fire: On the matter of the mind. New York: Basic Books.

Esquirol, Étienne 1845. Mental maladies; a treatise on insanity (original French edition 1838).

Ellis, E., Ames, M.A., 1987. Neurohormonal functioning and sexual orientation: a theory of homosexuality–heterosexuality. Psychol. Bull. 101.

Farrell MJ, Robertson IH. 2000. The automatic updating of egocentric spatial relationships and its impairment due to right posterior cortical lesions. Neuropsychologia.

Fasold O, von Bevern M, Kuhberg M, Ploner CJ, Vilringer A, Lempert T, Wenzel R. 2002. Human vestibular cortex as identified with caloric vestibular stimulation by functional magnetic resonance imaging. Neuroimage.

Falk, D. et al. Early hominid brain evolution: a new look at old endocasts. Journal of Human Evolution 38, (2000)

Fantoli, Annibale (2003). Galileo — For Copernicanism and the Church, 3rd English edition, tr. George V. Coyne, SJ. Vatican Observatory Publications, Notre Dame.

Farah, M.J. (1989), 'The neural basis of mental imagery', Trends in Neurosciences, 10.

Freud, S. (1905) "Three essays on the Theory of Sexuality"

"Form and Content in Totemism," American Anthropologist, Vol. 20, 1918.

Fiorini, M., Rosa, M.G.P., Gattass, R. & Rocha-Miranda, C.E. (1992), 'Dynamic surrounds of receptive fields in primate striate cortex: A physiological basis', Proceedings of the National Academy of Science 89.

Fodor, J.A. (1975), The Language of Thought (Cambridge, MA: Harvard University Press).

Frith, C.D. & Dolan, R.J. (1997), 'Abnormal beliefs: Delusions and memory', Paper presented at the May, 1997, Harvard Conference on Memory and Belief.

Finlay BL, Darlington RB (1995) Linked regularities in the development and evolution of mammalian brains. Science 268

Gazzaniga, M. S. (1985). The social brain. New York: Basic Books.

Gazzaniga, M.S. (1993), 'Brain mechanisms and conscious experience', Ciba Foundation Symposium, 174.

Greenspan, S. I. and S. G. Shanker (2004). The first idea: How symbols, language, and intelligence evolved from our early primate ancestors to modern humans. Cambridge, MA: Da Capo Press.

Goodwin GM, McCloskey DI, Matthews PBC. 1972. Proprioceptive illusions induced by muscle vibration: contribution by muscle spindles to perception? Science.

Good News Bible, New International Version

Geschwind N. "Behavioural changes in temporal lobe epilepsy". Psychol Med. 1979.

Greyson, Bruce. 2006. "Near-Death Experiences and Spirituality." Zygon: Journal of Religion and Science.

Grossman E, Donnelly M, Price R, Pickens D, Morgan V, Neighbor G, and others. 2000. Brain areas involved in perception of biological motion. J Cogn Neurosci.

Grüsser OJ, Landis T. 1991. The splitting of "I" and "me": heautoscopy and related phenomena. In: Visual agnosias and other disturbances of visual perception and cognition. Amsterdam: MacMillan.

Guru Granth Sahib -English Version 2012.

Gloor, P., Olivier, A., Quesney, L.F., Andermann, F., Horowitz, S. (1982), 'The role of the limbic system in experiential phenomena of temporal lobe epilepsy', Annals of Neurology, 12.

Gloor, P. (1992), 'Amygdala and temporal lobe epilepsy', in The Amygdala: Neurobiological Aspects of Emotion, Memory and Mental Dysfunction, ed J.P. Aggleton (New York: Wiley-Liss).

Grady, D. (1993), 'The vision thing: Mainly in the brain', Discover, June.

Green, R.E. A draft sequence of the Neandertal genome. Science 328

Gilbert SL, Dobyns WB, Lahn BT (2005) Genetic links between brain development and brain evolution. Nat Rev Genet 6

Gay, Volney, ed. Neuroscience and Religion. Plymouth, UK: Lexington Books, 2009.

Hall, Daniel, Keith Meador, and Harold Koenig. "Measuring Religiousness in Health Research: Review and Critique." Journal of Religion and Health 47, no. 2 (2008)

Harris, Sam, Jonas Kaplan, Ashley Curiel, Susan Bookheimer, Marco Iacoboni, and Mark Cohen. "The Neural Correlates of Religious and Nonreligious Belief." PLoS One 4, no. 10 (October 1, 2009)

Hood, Ralph, Peter Hill, and W. Paul Williamson. The Psychology of Religious Fundamentalism. New York: Guilford Press, 2005

Horgan, John. "The God Experiments." Discover Magazine 27, no. 12 (December 2006)

Halgren, E. (1992), 'Emotional neurophysiology of the amygdala within the context of human cognition', in The Amygdala: Neurobiological Aspects of Emotion, Memory and Mental

Dysfunction, ed J.P. Aggleton (New York: Wiley-Liss).

Harcourt-Smith, W. E. & L.C. Aiello. Fossils, feet and the evolution of human bipedal locomotion. Journal of Anatomy 204 (2004)

Horgan, J. (1994), 'Can science explain consciousness?', Scientific American, 271.

Humphrey, N. (1993), A History of the Mind (London: Vintage).

Hublin, J.J. The origin of Neanderthals. PNAS 45, (2009)

Henshilwood, C.S., Marean, C.W., 2003. The origin of modern human behavior: critique of the models and their test implications. Current Anthropology 44.

Heath, Thomas (1913). Aristarchus of Samos. Oxford: Clarendon Press.

Hoyle, Fred (1973). Nicolaus Copernicus.

Hilton, C.E. (Eds) From Biped to Strider: The Emergence of Modern Human Walking, Running, and Resource Transport. Kluwer Academic/Plenum, New York.

Haeusler, M., McHenry, H., 2004. Body proportions of Homo habilis reviewed. Journal of Human Evolution 46.

Hobbs, J. (2006). The origins and evolution of language: A plausible strong-AI account. In M. Arbibi (Ed.), Action to language via the mirror neuron system. Cambridge: Cambridge University Press.

Holloway RL, Broadfield DC, Yuan MS (2004) The human fossil record, vol 3, Brain endocasts: the paleo- neurological evidence. Wiley, New York

Holloway RL (1996) Evolution of the human brain. In: Lock A, Peters CR (eds) Handbook of human symbolic evolution. Oxford University Press, Oxford

Halligan PW, Fink GR, Marshal JC, Vallar G. 2003. Spatial cognition: evidence from visual neglect. Trends Cogn Sci.

Hécaen H, Ajuriaguerra J. 1952. L'Héautoscopie. In: Méconnassiances et hallucinations corporelles. Paris: Masson.

Hécaen H, Green A. 1957. Sur l'héautoscopie. Encephale.

Irwin HJ. 1985. Flight of mind: a psychological study of the out-of- body experience. Metuchen (NJ): Scarecrow Press.

Jackendoff, R. (1987), Consciousness and the Computational Mind (Cambridge, MA: The MIT Press).

Jaspinder Singh, The Sikh Gurus - Lives and Teachings: Spiritual Enlightenment Through Message Of Sikhism, Jawahar Publishers 2014.

Johansen, K. F.; Rosenmeier, H. (1998). A History of Ancient Philosophy: From the Beginnings to Augustine.

Kanizsa, G. (1979), Organization In Vision (New York: Praeger).

Kuypers HGJM (1958) Corticobulbar connections to the pons and lower brainstem in man. Brain 81

Kinsbourne, M. (1995), 'The intralaminar thalamic nucleii', Consciousness and Cognition, 4.

Kunimatsu, Y. et al. A new Late Miocene great ape from Kenya and its implications for the origins of African great apes and humans. PNAS USA 104. (2007)

King, W., 1864. The reputed fossil man of the Neanderthal. Quarterly Review of Science 1.

Kölmel HW. 1985. Complex visual hallucinations in the hemianopic field. J Neurol Neurosurg Psychiatry.

Kjaer, Troels, Camilla Bertelsen, Paola Piccini, David Brooks, Jorgen Alving, and Hans Lou. "Increased Dopamine Tone during Meditation-Induced Change of Consciousness." Cognitive Brain Research 13, no. 2 (April 2002)

Koenig, Harold. "Research on Religion, Spirituality, and Mental Health: A Review." Canadian Journal of Psychiatry 54, no. 5 (May 2009)

Koenig, Harold, ed. Handbook of Religion and Mental Health. San Diego, CA: Academic Press, 1998

Koestler, Arthur (1986) [1959]. The Sleepwalkers: A History of Man's Changing Vision of the Universe. Penguin Books.

Kuhn, Thomas S. (1957). The Copernican Revolution. Cambridge: Harvard University.

Lauglin, Charles, John McManus, and Eugene d'Aquili. Brain, Symbol, and Experience. 2nd ed. New York: Columbia University Press, 1992

Lakoff, G. and M. Johnson (1999). Philosophy in the flesh. Basic Books: New York.

LeDoux, J. E. (1996). The emotional brain. New York: Simon & Schuster.

LeVay, S., 1991. A difference in hypothalamic structure between heterosexual and homosexual men. Science 253.

Linton, Christopher M. (2004), From Eudoxus to Einstein—A History of Mathematical Astronomy, Cambridge: Cambridge University Press.

Litovitz TA, Penafiel M,Krause D,Zhang D,Mullins JM. The role of temporal sensing in bioelectromagnetic effects. Bioelectromagnetics 1997.

Lagace N, St-Pierre LS, Persinger MA. Attenuation of epilepsy-induced brain damage in the temporal cortices of rats by exposure to LTP-patterned magnetic fields. Neurosci Lett 2009.

Lackner JR. 1988. Some proprioceptive influences on the perceptual representation of body shape and orientation. Brain.

Lackner JR. 1992. Sense of body position in parabolic flight. Ann N Y Acad Sci.

Lawson, Russell M. (2004). Science in the Ancient World: An Encyclopedia. ABC-CLIO.

Leube DT, Knoblich G, Erb M, Grodd W, Bartels M, Kircher TT. 2003. The neural correlates of perceiving one's own movements. Neuroimage.

Lippman CW. 1953. Hallucination of physical duality in migraine. J Nerv Ment Dis.

Lobel E, Kleine J, Leroy-Wilig A. 1999. Functional MRI of galvanic vestibular stimulation. J Neurophysiol.

LeDoux, J.E. (1992), 'Emotion and the amygdala', in The Amygdala: Neurobiological Aspects of Emotion, Memory and Mental Dysfunction, ed J.P. Aggleton (New York: Wiley-Liss).

Le Gros Clark W.E., 1964. The fossil evidence for human evolution, 2nd ed. Chicago: University of Chicago Press.

Leakey, L.S.B., Tobias, P.V., Napier, J.R., 1964. A new species of the genus Homo from Olduvai Gorge. Nature 202.

Lama Surya Das, Awakening the Buddha Within: Tibetan Wisdom for the Western World, Broadway Books; Reprint edition 1998.

Makarec, K., & Persinger, M. A. (1985). Temporal lobe signs: Electroencephalographic validity and enhanced scores in special populations. Perceptual and Motor Skills, 60

Makarec, K., & Persinger, M. A. (1990). Electroencephalographic validation of a temporal lobe signs inventory in a normal population. Journal of Research in Personality, 24

Metzinger T. 2003. Being no one. Cambridge (MA): MIT Press.

Mittelstaedt H, Glasauer S. 1993. Illusions of verticality in weightlessness. Clin Invest.

Maryansky, A. (1996). African Ape social structure: A blue print for reconstructing early hominid structure. In J. Steel, S. Sherman (Eds.), The Archeology of Human Ancestry. London: Rutledge.

Massey, D. (2000). What I don't know about my field but wish I did. Annual Review of Sociology.

Massey, D. S. (2002). A brief history of human society: The origin and role of emotion in social life: 2001 presidential address. American Sociological Review.

McFadden, D., 1993. A masculinizing effect on the auditory systems of human females having male co-twins. Proc. Natl Acad. Sci. USA 90.

McFadden, D., Champlin, C.A., 2000. Comparison of auditory evoked potentials in heterosexual, homosexual and bisexual males and females. J. Assoc. Res. Otolaryngol. 1.

McFadden, D., Pasanen, E.G., 1998. Comparison of the auditory systems of heterosexuals and homosexuals: click-evoked otoacoustic emissions. Proc. Natl Acad. Sci. USA 95.

McFadden, D., Pasanen, E.G., 1999. Spontaneous otoacoustic emissions in heterosexuals, homosexuals and bisexuals. J. Acoust. Soc. Am. 105.

McMullen, Emerson Thomas, Galileo's condemnation: The real and complex story (Georgia Journal of Science, vol.61(2) 2003)

McMullin, Ernan, ed. (2005). The Church and Galileo. University of Notre Dame Press, Notre Dame.

Miller, B. D. (2007). Cultural anthropology, 4th ed. Boston: Allyn & Bacon.

Moody, Paul. 1975. Life after Life: The Investigation of a Phenomenon—Survival of Bodily Death. Atlanta: Mockingbird Books.

Mulligan BP, Hunter MD, Persinger MA. Effects of geomagnetic activity and atmospheric power variations on quantitative measures of brain activity: replication of the Azerbaijani studies. Adv Space Res 2010.

Martens, P.R. 1994. "Near-Death Experiences in Out-of-Hospital Cardiac Arrest Survivors. Meaningful Pheneomena or just Fantasy of Death?" Resuscitation.

Mounier, A., Marchal, F., Condemi, S. 2009. Is Homo heidelbergensis a distinct species? New insight on the Mauer mandible". Journal of Human Evolution 56.

McHenry, H., 1998. Body proportions in Australopithecus afarensis and A. africanus and the origin of the genus Homo. Journal of Human Evolution 35.

McHenry, H. M. Body size and proportions in early hominids. American Journal of Physical Anthropology 87 (1992)

McBrearty, S., Brooks, A., 2000. The revolution that wasn't: a new interpretation of the origin of modern humans. Journal of Human Evolution 39.

MacLean, P.D. (1990), The Triune Brain in Evolution (New York: Plenum Press).

MacKay, D.M. (1969), Information, Mechanism and Meaning (Cambridge, MA: The MIT Press).

Marr, D. (1982), Vision (San Francisco: Freeman). Medawar, P. (1969), Induction and Intuition in Scientific Thought (London: Methuen).

Milner, A.D. & Goodale, M.A. (1995), The Visual Brain In Action (Oxford: Oxford University Press).

McCullough, Michael, Ken Pargament, and Carl Thoresen, eds. Forgiveness: Theory, Practice, and Research. New York: The Guilford Press, 2000

Mecklenburger, Ralph. Our Religious Brains. Woodstock, VT: Jewish Lights Publishing, 2012

Moody, Raymond. Life after Life. New York: HarperCollins, 2001

Naskar, A. "The God Parasite: Revelation of Neuroscience", 2015

Naskar, A. "Your Own Neuron: A Tour of Psychic Brain", 2015

Naskar, A. "Homo: A Brief History of Consciousness", 2015

Naskar, A. "Autobiography of God: Biopsy of A Cognitive Reality", 2016

Newberg, Andrew, and Jeremy Iversen. "The Neural Basis of the Complex Mental Task of Meditation: Neurotransmitter and Neurochemical Considerations." Medical Hypotheses 61, no. 2 (2003)

Newberg, Andrew, and Mark Waldman. How God Changes Your Brain. New York: Ballantine, 2010

Newberg, Andrew, and Stephanie Newberg. "A Neuropsychological Perspective on Spiritual Development." In Handbook of Spiritual Development in Childhood and Adolescence, edited by Eugene Roehlkepartain, Pamela King,

Linda Wagener, and Peter Benson. London: Sage Publications, Inc., 2005

Newberg, Andrew, Nancy Wintering, Dharma Khalsa, Hannah Roggenkamp, and Mark Waldman. "Meditation Effects on Cognitive Function and Cerebral Blood Flow in Subjects with Memory Loss: A Preliminary Study." Journal of Alzheimer's Disease 20, no. 2 (2010)

Newberg, Andrew, Nancy Wintering, Mark Waldman, Daniel Amen, Dharma Khalsa, and Alavi Alavi. "Cerebral Blood Flow Differences between Long- Term Meditators and Non-Meditators." Consciousness and Cognition 19, no. 4 (2010)

Newberg, A. "Cerebral blood flow changes associated with different meditation practices and perceived depth of meditation" Psychiatry Research: Neuroimaging 2010.

Neisser U. 1988. The five kinds of self-knowledge. Phil Psychol.

Nash, M. (1995), 'Glimpses of the mind', Time.

Nimchinsky EA, Gilissen E, Allman JM, Perl DP, Erwin JM and Hof PR (1999) A neuronal morphologic type unique to humans and great apes. Proc Natl Acad Sci USA 96

Nussbaum, Alexander (January–April 2002). "Creationism and geocentrism among Orthodox Jewish scientists". Reports of the National Center for Science Education.

Persinger, "'I would kill in God's name' role of sex, weekly church attendance, report of a religious experience and limbic lability" Perceptual and Motor Skills 1997.

Persinger "Experimental simulation of the God experience" Neurotheology 2003.

Persinger, M. A. (1993b). Personality changes following brain injury as a grief response to the loss of sense of self: Phenomenological themes as indices of local lability and neurocognitive restructuring as psycho- therapy. Psychological Reports, 72

Persinger, Corradini, Clement, Keaney, et al "Neurotheology and its convergence with neuroquantology" NeuroQuantology 2010.

Persinger, Koren and St-Pierre "The electromagnetic induction of mystical and altered states within the laboratory" Journal of Consciousness Exploration and Research 2010.

Persinger "Case report: A prototypical spontaneous 'sensed presence' of a sentient being and concomitant electroencephalographic activity in the clinical laboratory" Neurocase 2008.

Persinger and Saroka "Potential production of Hughlings Jackson's "parasitic consciousness" by physiologically-patterned weak transcerebral magnetic fields: QEEG and source localization" Epilepsy & Behavior 28 (2013).

Persinger. "The neuropsychiatry of paranormal experiences". J Neuropsychiatry Clin Neurosci 2001.

Persinger. "Neuropsychological bases of god beliefs", New York: Praeger, 1987

Persinger. "Temporal lobe epileptic signs and correlative behaviors displayed by normal populations", Journal of General Psychology, 1986

Persinger "Experimental Facilitation of the Sensed Presence: Possible Intercalation between the Hemispheres Induced by Complex Magnetic Fields" Journal of Nervous and Mental Disease 2002.

Paré, D. & Llinás, R. (1995), 'Conscious and preconscious processes as seen from the standpoint of sleep-waking cycle neurophysiology', Neuropsychologia, 33.

Penfield, W.P. & Jasper, H. (1954), Epilepsy and the Functional Anatomy of the Human Brain (Boston, MA: Little, Brown & Co.).

Penfield, W.P. & Perot, P. (1963), 'The brain's record of auditory and visual experience: a final summary and discussion', Brain, 86.

Penrose, R. (1994), Shadows of the Mind (Oxford: Oxford University Press).

Penrose, R. (1989), The Emperor's New Mind: Concerning Computers, Minds and The Laws of Physics (Oxford: Oxford University Press).

Posner, M.I. & Raichle, M.E. (1994), Frames of Mind (New York: Scientific American Library).

Preuss TM, Caceres M, Oldham MC, Geschwind DH (2004) Human brain evolution: insights from micro- arrays. Nat Rev Genet 5

Peter Furst, Flesh of the Gods: The Ritual Use of Hallucinogens (New York: Waveland, 1990)

Paul Devereux, The Long Trip: A Prehistory of Psychedelia (New York: Penguin, 1997)

Pickford, M., Senut, B., 2001. 'Millennium Ancestor', a 6-million-year-old bipedal hominid from Kenya - Recent discoveries push back human origins by 1.5 million years. South African Journal of Science 97.

Paloutzian, Ray, and Crystal Park, eds. The Handbook of the Psychology of Religion. New York: Guilford Publications, Inc., 2005

Peter Stafford, Psychedelics Encyclopedia (Berkeley, CA: Ronin Press, 1992)

Rabinowitz, Avi (1987). "EgoCentrism and GeoCentrism; Human Significance and Existential Despair; Bible and Science; Fundamentalism and Skepticalism". Science & Religion.

Rene Descartes, "The Inter-Relation of Soul and Body," in The Way of Philosophy, edited by P. Wheelright (New York: Odyssey, 1954)

Richardson, K. (1999). The making of intelligence. London: Phoenix.

Ratnasuriya, R.H. "Joan of Arc, creative psychopath: is there another explanation?" Journal of The Royal Society of Medicine 1986.

Ramachandran, V.S. (1993), 'Filling in gaps in logic: Some comments on Dennett', Consciousness and Cognition, 2.

Ramachandran,V.S.(1995a),'Filling in gaps in logic: Repy to Durginetal.',Perception, 24.

Ramachandran, V.S. (1995b), 'Perceptual correlates of neural plasticity', in Early Vision and Beyond, ed. T.V. Papathomas, C. Chubb, A. Gorea and E. Kowler (Cambridge, MA: The MIT Press).

Ramachandran, V.S. & Gregory, R.L. (1991), 'Perceptual filling in of artificially induced scotomas in human vision', Nature, 350.

Ramachandran, V.S. and Blakeslee, S. (1999), Phantoms in the Brain: Probing the Mysteries of the Human Mind (William Morrow Paperbacks)

Rahman, Q., Abrahams, S., Wilson, G.D., 2003a. Sexual orientation related differences in verbal fluency. Neuropsychology 17.

Rilling, J. K. (2006). Human and nonhuman primate brains: Are they allometrically scaled versions of the same design? Evolutionary Anthropology, 15.

Relethford, J. H. Genetic evidence and the modern human origins debate. Heredity 100, (2008)

Rightmire, G. P. Out of Africa: modern human origins special feature: middle and later Pleistocene hominins in Africa and Southwest Asia. PNAS USA 106, (2009)

Rightmire, G.P. Homo in the Middle Pleistocene: Hypodigms, variation, and species recognition. Evolutionary Anthropology 17, 8-21 (2008)

Roebroeks, W. & P. Villa. On the earliest evidence for habitual use of fire in Europe. PNAS USA (2011)

Roth, G. and Dicke, U. Evolution of the brain and intelligence, TRENDS in Cognitive Sciences Vol. 9, No. 5, 2005

Rightmire, G.P., 1998. Human evolution in the Middle Pleistocene: the role of Homo heidelbergensis. Evolutionary Anthropology 6.

Ruby P, Decety J. 2001. Effect of subjective perspective taking during simulation of action: a PET investigation of agency. Nat Neurosci.

Ronald Siegel, Intoxication: Life in Pursuit of Artificial Paradise (New York: EP Dutton, 1989)

R. Gordon Wasson, Carl A. P. Ruck, and Stella Krammrisch, Persephone's Quest: Entheogens and the Origins of Religion (New Haven, CT: Yale Univer- sity Press, 1988).

Richard E. Schultes and Albert Hofmann, Plants of the Gods (New York: McGraw Hill, 1979)

Richard E. Schultes and Albert Hofmann, The Botany and Chemistry of Hallucinogens, 2nd ed. (Springfield, IL: Charles C. Thomas, 1980); and Jonathan Ott, Pharmacotheon (Kennewick, WA: Natural Products Co., 1993).

N. Pahnke, Albert A. Kurland, Sanford Unger, Charles Savage, and Stanislav Grof, "The Experimental Use of Psychedelic (LSD) Psychotherapy," Journal of the American Medical Association 212 (1970)

Rick J. Strassman, "Adverse Reactions to Psychedelic Drugs. A Review of the Literature," Journal of Nervous and Mental Disease 172 (1984)

R. H. F. Manske, "A Synthesis of the Methyl-Tryptamines and Some Derivatives," Canadian Journal of Research 5 (1931)

Steven A. Barker, John A. Monti, and Samuel T. Christian, "N,N-Dimethyltryptamine: An Endogenous Hallucinogen," International Review of Neurobiology 22 (1981)

Schultes, Richard, Albert, Hofmann, and Christian Rätsch. Plants of the Gods: Their Sacred, Healing, and Hallucinogenic Powers. Rochester, VT: Healing Arts Press, 2001

Shermer, Michael. The Believing Brain: From Ghosts and Gods to Politics and Conspiracies— How We Construct Beliefs and Reinforce Them as Truths. New York: Times Books, 2011

Sawer, G. and Deak, V. (2007). The last human. New York: Peter N. Nevraumont Publication – Yale University Press.

Small, D. (2008). On the deep history of the brain. Berkeley: University of California Press.

Sheils D. 1978. A cross-cultural study of beliefs in out-of-the-body experiences, waking and sleeping. J Soc Psych Res.

Smith BH. 1960. Vestibular disturbances in epilepsy. Neurol.

Srimad-Bhagavatam, The Bhaktivedanta Book Trust 2012.

Streeter, Chris, J. Eric Jensen, Ruth Perlmutter, Howard Cabral, Hua Tian, Devin Terhune, Domenic Ciraulo, and Perry Renshaw. "Yoga Asana Sessions Increase Brain GABA Levels: A Pilot Study." Journal of Alternative and Complementary Medicine 13, no. 4 (May 2007)

Strassman, R. "DMT: The Spirit Molecule" 2001.

Slater E, Beard AW. The schizophrenia-like psychoses of epilepsy. Br J Psychiatry 1963.

Searle, John R. (1980), 'Minds, brains, and programs', Behavioral and Brain Sciences, 3.

Searle, John R. (1992), The Rediscovery of the Mind (Cambridge, MA: The MIT Press).

Semendeferi K, Lu A, Schenker N, Damasio H (2002) Humans and great apes share a large frontal cortex. Nat Neurosci 5

Strauss, F., Risser, A. & Jones, M.W. (1982), 'Fear responses in patients with epilepsy', Archives of Neurology, 39.

Schoetensack, O., 1908. Der Unterkiefer des Homo heidelbergensis aus den Sanden von Mauer bei Heidelberg. Leipzig: Wilhelm Engelmann.

Senut, B. et al. First hominid from the Miocene (Lukeino Formation, Kenya). C. R. Acad. Sci. Paris, Sciences de la Senut, B., Pickford, M., Gommery, D., Mein, P., Cheboi, K., Coppens, Y., 2001. First hominid from the Miocene (Lukeino Formation, Kenya). Comptes Rendus De L Academie Des Sciences Serie Ii Fascicule a-Sciences De La Terre Et Des Planetes 332.

Sherwood CC, Broadfield DC,Gannon PJ,Holloway RL, Hof PR (2003) Variability of Brocas area homologue in African great apes: implications for language evolution. Anat Rec 71A

Stringer, C.B., Finlayson, J.C., Barton, R.N.E, Fernández-Jalvo, Y., Cáceres, I., Sabin, R.C., Rhodes, E.J., Currant, A.P., Rodríguez-Vidal, J., Giles-Pacheco, F., Riquelme-Cantal, J.A., 2008. Neanderthal exploitation of marine mammals in Gibraltar. Proceedings of the National Academy of Sciences USA 105.

Shipman, P., 2008. Separating "us" from "them": Neanderthal and modern human behavior.

Proceedings of the National Academy of Sciences USA 105.

Schmitt, D., Churchill, S., 2003. Experimental evidence concerning spear use in Neandertals and early modern humans. Journal of Archaeological Science 30.

Shukla, Kripa Shankar. Aryabhata: Indian Mathematician and Astronomer. New Delhi: Indian National Science Academy, 1976.

Sutherland, N.S. (1989), The International Dictionary of Psychology (New York: Continuum).

Sawer, G. and Deak, V. (2007). The last human. New York: Peter N. Nevraumont Publication – Yale University Press.

Small, D. (2008). On the deep history of the brain. Berkeley: University of California Press.

Terence McKenna, Food of the Gods (New York: Bantam, 1993)

Turner, B. (2000a). Embodied ethnography. Doing culture. Social Anthropology.

Turner, J. H. (2000b). On the origins of human emotions: A sociological inquiry into the evolution of human affect. Stanford, California: Stanford University Press.

The Quran, Translated by Muhammad Abdel Haleem, Oxford University Press 2008.

The Upanishads (Classic of Indian Spirituality), Nilgiri Press 2009.

The Upanishads (Penguin Classics), Penguin Classics; Reissue edition 1965.

The Bhagavad Gita (Classics of Indian Spirituality), Nilgiri Press 2007.

The Mahabharata (Penguin Classics), Penguin Classics; Abridged edition 2009.

The Book of Mormon: Another Testament of Jesus Christ, Church of Jesus Christ of Latter Day Saints 1981.

Tiller, S. G., & Persinger, M. A. (2002). Geophysical variables and behavior: XCVII. Increased proportions of left-sided sense of presence induced experimentally by right hemispheric application of specific (frequency-modulated) complex magnetic fields. Perceptual and Motor Skills, 94

Timothy Leary, "Programmed Communication During Experiences with DMT," Psychedelic Review 8 (1966)

Tononi, G., & Edelman, G. E. (1998). Consciousness and complexity. Science, 282

Tulving, E. (1983), Elements of Episodic Memory (Oxford: Clarendon Press).

Trimble, M.R. (1992), 'The Gastaut-Geschwind syndrome', in The Temporal Lobes and the Limbic System, ed. M.R. Trimble and T.G. Bolwig (Petersfield: Wrightson Biomedical Publishing Ltd.).

Trinkhaus, E., 1985. Pathology and the posture of the La Chappelle-aux-Saints Neanderthal. American Journal of Physical Anthropology 67.

Trinkaus, E., Shipman, P., 1993. The Neanderthals: Changing the Image of Mankind. Knopf: New York.

Ungar, P.S., Grine, F.F., Teaford, M.F., 2006. Diet in early Homo: a review of the evidence and a new model of adaptive versatility. Annual Review of Anthropology 35.

Warner, Bill. 2010, Sharia Law For The Non-Muslim, Center For The Study of Political Islam

Westfall, Richard S. (1989). Essays on the Trial of Galileo. Vatican City: Vatican Observatory.

White, Michael (2007). Galileo antichrist. London: Wiedenfeld & Nicholson.

Whinnery, J.E. 1997. "Psychophysiologic Correlates of Unconsciousness and near- death experiences." Journal of Near-Death Studies.

Waxman, S.G. & Geschwind, N. (1975), 'The interictal behavior syndrome of temporal lobe epilepsy', Archives of General Psychiatry, 32

Wilson, David Sloan. Darwin's Cathedral: Evolution, Religion, and the Nature of Society. Chicago: Chicago University Press, 2003

Wade Davis and Andrew T. Weil, "Identity of a New World Psychoactive Toad," Ancient Mesoamerican (1988)

Willis W. Harman, Robert H. McKim, Robert E. Mogar, James Fadiman, and Myron J. Stolaroff, "Psychedelic Agents in Creative Problem-Solving: A Pilot Study," Psychological Reports 19 (1966)

Wong, K., 2010. Fossils of our family. Scientific American June 2010.

Wray,A.(1998)."Protolanguage as a holistic system for social interaction, "Language & Communication 18.

Young, N. M. et al. The phylogenetic position of Morotopithecus. Journal of Human Evolution 46, (2004)

Zeki, S.M. (1978), 'Functional specialisation in the visual cortex of the rhesus monkey', Nature, 274.

Zeki, S.M. (1993), A Vision of the Brain (Oxford: Oxford University Press).

Zacks JM, Ollinger JM, Sheridan MA, Tversky B. 2002. A parametric study of mental spatial transformations of bodies. Neuroimage.

* * *

www.ingramcontent.com/pod-product-compliance
Lightning Source LLC
Chambersburg PA
CBHW030935180526
45163CB00002B/573